Introduction to the
FISHERY SCIENCES

WILLIAM F. ROYCE

COLLEGE OF FISHERIES
UNIVERSITY OF WASHINGTON
SEATTLE, WASHINGTON

ACADEMIC PRESS   New York   San Francisco   London

A Subsidiary of Harcourt Brace Jovanovich, Publishers

ACADEMIC PRESS, INC.
111 Fifth Avenue, New York, New York 10003

United Kingdom Edition published by
ACADEMIC PRESS, INC. (LONDON) LTD.
24/28 Oval Road, London NW1 7DD

LIBRARY OF CONGRESS CATALOG CARD NUMBER: 79-180795

PRINTED IN THE UNITED STATES OF AMERICA

Introduction to
The Fishery Sciences

In an Egyptian village near the Nile River, about 2500 B.C., the peasants caught and prepared fish for food while a leader became renowned for his prowess at spearfishing. Redrawn from Percy E. Newberry (1893) "Beni Hasan," Part II, Archaeological Survey of Egypt, Kegan Paul, Trench, Trubner & Co., London.

Contents

9   Fishery Resource Management

Preface

This text is intended to be used by students who want a broad introduction to fishery science either as background for further studies that will enable them to enter the profession or as a survey of a natural resource area that will inform them about the public issues. It considers not only fish but the aquatic mammals, mollusks, crustaceans, and other organisms that are commonly included in the fisheries. It covers not only the biology of the resource organisms but their ecology and some of the legal, social, and political aspects of their use as well.

The early fishery scientists received their education and training in a variety of biological disciplines. Many were naturalists who concerned themselves primarily with identification and nomenclature of fish, a speciality that is more properly called ichthyology. Others were trained especially in the sciences fundamental to aquaculture. Still others were trained in limnology, oceanography, or marine biology. In every case the universities and colleges emphasized broad training in basic physical and biological sciences.

The first special fishery courses in North American universities were introduced about 1912 at Cornell University (aquaculture) and about 1919 at the University of Washington (fish technology and aquaculture). These were the principal programs until the late 1930's when a number of universities initiated programs in fishery biology and fishery management in response to the growing concern about the decline in fish populations. The programs proliferated until there were 107 universities offering fishery science-related programs in 1967, although only 20 were giving courses in all three areas of fishery biology, fishery research, and fishery management. Recently, several universities have combined the fishery programs with other resource or environmental management programs.

Employment of professional fishery scientists has grown rapidly since 1940. In the United States about 5500 positions requiring at least a B. S. degree had been established by 1970. Of these, about 1757 were with the federal government, 2010 with state governments, 1100 with the private sector, and 600 with universities. About 60% of these positions were with fishery agencies, of which about three-fourths were involved predominantly with recreational fisheries and about one-fourth with commercial fisheries. Outside the fishery agencies the principal government employers were agencies concerned with the management of forests, lands, parks, or water quality. The positions in the private sector were principally with industries having aquatic environmental problems, the commercial fishing industry, aquaculture, aquaria, or conservation organizations.

Fishery professions, other than biology, have emerged recently. A few positions are available as fishery engineers, fishery economists, or fishery lawyers, but their numbers are rapidly growing as people seek solutions to the many nonbiological fishery problems.

The professional fishery science positions are numerous in the developed countries with a strong scientific tradition, although a comprehensive count is not available. In addition, most of the developing countries have recognized the need for professionals in their fishery agencies, and positions requiring them are being established

Thus, at least 10,000 positions exist for fishery scientists, a large proportion of which are involved with the management and use of the fishery resources. Those filling these positions comprise an identifiable profession of people who face similar technical and social problems, who communicate among themselves to a major extent, who have developed standards of professional competence, and who need some common education in fisheries science.

I am heavily indebted to many colleagues who have helped by a review of chapters or a discussion of issues. Included are D. Lee Alverson, Robert L. Burgner, Douglas G. Chapman, Allan C. DeLacy, Lauren R. Donaldson, Richard H. Fleming, Donald W. Hagen, John E. Halver, G. Ivor Jones, Hiroshi Kasahara, Loyd A. Royal, Lynwood S. Smith, Frieda B. Taub, Arthur D. Welander, Henry O. Wendler, Richard R. Whitney, and Norman J. Wilimovsky.

Others who have assisted greatly with editing, typing, and illustrating are Remedios W. Moore, Linda A. Strickler, and Diana A. Wetherall.

WILLIAM F. ROYCE

**Introduction to
The Fishery Sciences**

The Challenge to Fishery Scientists

The use of aquatic resources has been a challenge to man throughout his existence on earth. The first men who lived on the shore of a sea, river, or lake found food in the animals, fibers in the plants, and ornaments in the shells. They gradually learned to catch fish in traps, to make hooks from bone and metal, and to twist fibers into lines and nets. They learned to use reeds and wood to build rafts that would carry them along the shore to find better fishing and carry their catch home. As they began to farm, they discovered that they could raise fish in ponds or keep them in enclosures alongshore and thus have fish available when needed. They learned to make maps and bigger boats, and to navigate; thus they were able to fish in waters that were beyond the sight of land. They learned to preserve fish by drying them in the sun, and by smoking or salting them; but the catch had to be brought ashore within a few hours. They also had fun spearing or angling and became known for their prowess. However, they never completely mastered the resources of the sea. Some of its inhabitants remained monsters too large or fierce to overcome, and most were beyond their reach or power. The legends continued to grow that more and bigger fish remained in the sea than had ever been caught.

Today man can reach the deepest part of the ocean with a line; he can even visit it personally in a vehicle, he can fish easily in every ocean of the world and bring the fish home in good condition and he can subdue the largest animals on earth, the great whales. He has discovered that practically all of the waters that are not frozen or are not more than 50° C contain an intricate web of life in enormous variety from bacteria to whales that is sustained as an organic production system by energy from the sun. Because water covers nearly three-fourths of the earth, it receives much more energy than the land and is therefore a larger reservoir of energy and organic material.

Only part of the material is used by man: some of the larger fish, mollusks, crustaceans, mammals, and plants. The preponderance of the living material is comprised of the tiny forms, the bacteria, protozoa, one-celled plants called "diatoms," and minute crustaceans and other tiny animals. The plants convert nutrients and the energy of the sun into a form which sustains the larger organisms that a man may use.

Understanding this organic production system of energy, water, plants, and animals is a task for scientists from many disciplines. Physicists, chemists, geologists, and biologists of many kinds all play parts in the great scientific investigations. They comprise a host of basic researchers who seek knowledge for its own sake, and who unravel the mysteries of the waters and living things.

The fishery scientist is concerned with the things that are useful to man. *Fishery science* is the scientific study of the use of the living resources of the waters. Part of fishery science is concerned with the biological, physical, and chemical aspects of the process of organic production; part, with the distribution and abundance of resources; part, with the effects of fishing. It is an applied science, including study directed at basic understanding as well as study designed to provide a background for decisions.

1.1 Recent Trends in Foodfish Production

When the weary people of the world began to rebuild their countries and economies after World War II, they looked first to their supply of food. Those who had been fishermen went back to sea if they could find a ship. By 1948 they were operating enough old or makeshift vessels to bring the world production of fish back to the prewar level of about 20 million tons live weight per year. In 1948 they caught about 17.1 million tons from the sea and 2.5 million tons from fresh water.

The old ways of fishing were not followed for long because the wartime navies had brought many changes in man's knowledge of the ocean and in the equipment of his vessels. The oceans had been studied as never before so that submarines could be found and caught, a problem fundamentally the same as finding and catching fish. The current system of the oceans and the temperature structure were much better known. Ingenious acoustical equipment had been devised to probe beyond the limits of vision. Accurate new navigational systems and instruments were in common use. New synthetic fibers were replacing hemp in ropes and lines. Moreover, many men had gone to sea during the war, learned to live with it, and learned about the new equipment.

Soon people responded to the need for food with the new vessels which were designed to fish farther from port, navigate with the new equipment, find fish with sonar, catch fish with synthetic nets, dress the fish with new machines, and freeze the fish on board. All of these things did not happen immediately; mistakes were made, some ideas were discarded, and some gear modified.

But ingenious and persistent men prevailed, and a revolution in fishing developed. The new stern trawlers were as much of an advance over the trawlers of the 1930's as the reaper was over hand-threshing equipment. New fibers for nets that did not rot and a new power block for retrieving purse seines made possible as much of an improvement over the earlier purse seiners as a moldboard plow over a wooden stick. New sonar instruments and gear that could catch schools of fish midway between surface and bottom increased the productivity of herring and cod fishermen as much as the productivity of new varieties of corn surpassed the productivity of the old. New filleting, packaging, and freezing equipment that could be taken to sea made possible landing the catch in perfect condition, ready for market, and removed the worry about spoilage, as similar processing machinery on land had done for agricultural products.

The new fleets manned by skillful and daring men ranged all of the seas. Vessels were able to fish the North Pacific one month and the South Atlantic 2 months later. When catches were disappointing, they could move to new grounds—perhaps to another ocean. All of the known fish stocks in the world could now be fished.

The results of this new technology were soon apparent. The total production reached the prewar level of about 20 million tons in 1948; it exceeded 30 million tons in 1956, only 8 years later. It increased by 10 million tons in the next 5 years and another 10 million tons in the next 3 years. The freshwater production increased from 2.5 million tons in 1948 to 7.2 million tons in 1965, and the sea fisheries production increased from 14.8 million tons to 45.2 million tons—rates of growth of about 6% per year.

The capacities of the new fleets and the increases in catch brought concern for the welfare of the stocks of fish. None of the new fleets was restrained by law. Many of them did not furnish statistics on the amount of catch, the area of operation, and the fishing effort; they merely provided statistics on the total landed in a home port. Their mission was to provide food immediately for hungry people and their efforts brought constantly greater total catches. But the trend in some fisheries was ominous: the catches either leveled off or declined regardless of the amount of fishing. On the basis of biological evidence it was shown that stocks of the Arctic cod and haddock off northwest Europe, the Newfoundland fisheries, the yellowfin tuna of the eastern tropical Pacific, and the anchovy of Peru and Chile were producing near their maximum sustainable yields; yet based on the same evidence it could be shown that some could produce more with less fishing. The Antarctic whale fishery had almost disappeared plainly as a result of overfishing. New stocks of bottom fish off West Africa that had not been fished prior to 1950 were discovered, fished, reduced to an uneconomic level of abundance, and largely abandoned. More and more the question was asked: What are the limits?

1.2 The Need for Food

By the middle 1960's, because of the discouraging trends in population and food supply from the land, it was the hope of many people that the oceans would produce still greater amounts of food. Agricultural production had also increased after the war under the insistent demand for food and with the help of new technology, but not as much as fishery production had, while the population had increased even more. The overall index of agricultural production *per person* changed from 95 around 1950 to 107 in 1960/1961; thus the calorie intake per person increased by about 12% during the decade. But no further improvement came, and in 1965/1966 the index was 106. On the other hand, the comparable indexes *per person* for fish production were 92 around 1950, 106 in 1960/1961, and 118 in 1965/1966. The gain in fish production did little to compensate for the lag in agricultural production because the fisheries supplied only a small fraction of the total food. Good figures are not available, but fish probably supplied the world with about one-third as much food as meat from swine, cattle, sheep, and poultry.

The leveling off of the average supply of food per person was bad enough, but the situation of many individuals was even worse. The overall average concealed the figures that showed the population increases in the developed countries running behind the rate of increase in food production and ahead in the underdeveloped countries. The rich were better fed and the poor were more poorly fed. Even within the underdeveloped countries much of the increase in amount of better food was going to the small proportion of people with higher incomes. In addition, the overall production figures reflected the trend, not the total supply. The latter depended in part on entry into and withdrawal of food from storage; during the early 1960's the stocks of stored food were greatly reduced, especially in North America.

Another serious aspect of the shortage was the distribution of the types of food. The diet of people in the underdeveloped countries was high in carbohydrates, low in fats and protein. The last is the scarcest and most expensive of foodstuffs, and a large proportion of the people in the world have never had the amount or quality of proteins they need. The minimum amount required per person per day is estimated as at least 60 g. The quality is reckoned according to the proportion of animal proteins because only animal proteins (meat, milk, eggs, poultry, fish) supply all of the amino acids required. Some plant proteins, such as those from peas and beans, supply more essential amino acids than others, so the proportion of animal proteins required depends on the source of other proteins in the diet. But a minimum of 17 g of animal proteins is a standard, for which up to 10 g of protein from peas and beans may be substituted.

The consequences of a scanty diet, especially a lack of proteins of the right kind, are most serious among children. Newly weaned children in many countries receive only a starchy gruel that may be low in calories as well as protein. This deficiency leads to the disease called "kwashiorkor," which, in its severe form, is characterized by the failure to grow, edema, damage to numerous body organs, and lowered resistance to infection. Malnutrition is widespread among children in most of tropical Asia, Africa, and Latin America.

Thus, with a widespread shortage of protein, a static and probably soon to worsen food situation, the fisheries offer hope to people who are looking everywhere for food. They provide a high protein food with an excellent blend of essential amino acids; the production, although still small in relation to world food supply, has been responding to new technology about twice as fast as agriculture has responded. The central and urgent question posed to the fishery scientist is, therefore, how much food can be taken from the waters on a sustained basis?

1.3 The Need for Outdoor Recreation

As men leave their farms to live in ever more crowded cities, they retain a love of nature, a need for the peace and quiet of the forest, and an urge to meet the challenges of the wild. They may stroll through a city park, swim on an ocean beach, camp in a wilderness, or drive through scenic mountains. The scale of activity may be suitable for the elderly or the most active. To all, it offers a revitalization—escape from the irritations and pressures of the city.

A surprisingly large proportion of the people go fishing, either as a primary objective or as a supplement to boating, hiking, or camping. In the United States about one-fourth of the adults go fishing each year. They may seek solitude, caring little about the catch, or seek limits and trophies to satisfy the urge to compete. They may travel by city bus to a pier, or by private plane to a remote lake. The demand is widespread through all parts of the country, among people of all levels of income and all kinds of employment; and it is increasing as people can travel farther more easily. The total recreational catch by anglers is larger than the commercial catch in most freshwater areas and in some marine areas near centers of population.

In other countries the demand for recreational fishing is also increasing. Some anglers seek trout that have been stocked in mountain streams around the world. Others are discovering the huge freshwater fish of the Nile and Amazon or the marlins and tunas off the coast of tropical countries. These anglers are an important fraction of the tourists, whose money is sought by all countries.

Because anglers seek escape from the cities and because their fishing is frequently combined with other activities such as boating and hiking in which enjoyment of the environment plays a large part, they are especially sensitive to the condition of the environment. They demand clean waters without the stench or filth of pollution.

Anglers in the United States are numerous enough to be a potent political force. Long ago they requested artificially reared fish to augment the stocks of the natural waters and got major programs of hatchery operation. Sports fishermen so outnumber commercial fishermen (about 350 to 1) in all states that any conflict between sports and commercial fishermen is usually resolved in favor of the sportsmen; recreational use of fish stocks is declared in many official policy statements to be the primary use. The anglers' concern about pollution has been a primary force in a major shift of the national view of pollution from a danger to public health to a danger to the environment.

1.4 Common Property Resources

One of the older legal principles in most parts of the world is that wild fish in national waters are the property of the state. Another is that the fish in the high seas are the property of no one. In either case the fish usually become the property of the person who catches them. A corollary principle that fishing is open to all who want to fish, pertains in most countries. Exceptions are made of course for resources under cultivation, but even these may be closely controlled by the state.

These principles have consequences that affect profoundly the management of all wild aquatic resources. One consequence is that no individual will benefit proportionately from any unilateral restraint that he practices in fishing because if he liberates or avoids catching fish everyone else has the right to catch them. Another consequence interferes with the maintenance of economic efficiency in older fisheries. If restrictions are needed for the maintenance of production from a fishery resource, the government must reduce the number of operators or reduce the efficiency of all fishermen. The latter course is commonly followed because of strong public feeling about everyone's right to fish. The result in the older food fisheries is lack of net profit for most of the fishermen.

1.5 Conservation

The management of the aquatic resources in the United States and many other countries of the world is part of the conservation of all natural resources. The movement is derived from a basic love of nature and a feeling of security in the animals, the forests, and the land. It is one of the more powerful emotional and political forces in the United States, a force which always asserts the highest moral principles and is always suspicious of the profitable use of resources. In the minds of many, conservation means preservation; to others it may mean wise use or official supervision; in any case, most people feel that the natural resources belong to them and must be protected.

A clarification of the meaning of conservation of the living resources of the sea emerged from the International Conference on the Law of the Sea held in Geneva in 1958. It was defined as follows: ". . . the aggregate of the measures rendering possible the optimum sustainable yield from those resources so as to secure a maximum supply of food and other marine products. Conservation programmes should be formulated with a view to securing in the first place a supply of food for human consumption."

This definition plainly means enhancing the production from underused resources as well as restraining the fishing on overused resources. Unfortunately, fish that die naturally are not as obvious as fruits left to rot on a tree, and there is little public recognition that unused living aquatic resources are wasted. The instinct of most conservationists is to preserve them.

1.6 The Tasks of Fishery Scientists

Every country in the world will be confronting the problem of a scanty food supply for an expanding population. Most countries will have people who also demand more recreational facilities for citizens and tourists to use. The sea will provide the great majority of the opportunities for expanding fisheries but in addition every body of fresh water either existing or artificial will furnish some resources. The expansion of fish production from the wild stocks in public waters will be carried as far as possible, but an increasing proportion of fish will be supplied by aquaculture, either for food or recreation, and either public or private. Scientists will be asked to participate in three broad areas of work.

1.6.1 MANAGING THE FISHING ON PUBLIC RESOURCES

The first scientific task is to describe the stocks of wild animals and plants with respect to their abundance, migrations, distribution, behavior, size and age composition, rate of growth, and rate of mortality. A second task, closely related and usually simultaneous, is to describe the environmental factors that either limit or augment the stocks such as temperature, currents, bottom types, dissolved materials, and fertility of the water. A third task, also closely related and conducted simultaneously, is to describe the kind, distribution, and costs of fishing. Each of the three tasks contributes vital information to the final tasks of estimating the relation of the yield from the stocks to the amount of fishing and to natural events. All tasks are conducted with close integration of plans and operations.

The findings of these studies are used for one of two courses of action; either to encourage harvest of underused resources or to restrain harvest of overused resources. In either case the objective of the research and the subsequent decisions is to obtain maximum sustainable yield in either physical or economic units.

1.6.2 PROTECTING AND ENHANCING THE ENVIRONMENT

The demand for water is at least as insistent as the demand for food and recreation. Bodies of water serve many human needs, some of them in conflict with the needs of the living resources. Water may be diverted to supply cities, create power, provide transportation, supply industry, or irrigate lands, leaving behind streams much smaller than before and blocked by dams, which may interfere with migratory fish. Smaller streams may be eliminated completely by diversion through pipes; marshes and estuaries may be filled to create new land. The water in the streams may be used to dilute and degrade waste and be so changed that fish cannot live. Even the living resources of the sea may be endangered by pesticides and radioactive wastes.

Because the living resources are usually public, the fishery agencies commonly become advocates for the resources and the aquatic environment. They call on their scientists to estimate the probable effects of changed environments on the stocks. If damage is expected, the scientists may be asked to suggest either alterations in the proposed structures or compensatory facilities.

On the other hand, new bodies of water are being rapidly created because whenever water is diverted it must be transported and stored. Commonly the new reservoirs and occasionally the new canals are usable for fish production. In some cases the new reservoirs may be more promising than natural waters because unwanted species can be kept out. Fishery scientists may be asked to share in the planning in order to maximize the utility of the project.

Some environmental changes can be made for the enhancement of production of wild fish from natural waters. Shelters can be added to the shallow parts of the sea, to inland lakes, to streams. Obstructions can be removed so that there is free migration of fish; undesirable fish can be removed; fertility of the water can be augmented. Scientists will be asked to devise methods for further enhancing the production and to evaluate the benefits from alternative proposals for modification.

1.6.3 AQUACULTURE

Occasionally it is suggested that a solution to the food shortage is to farm the sea. Aquaculture is still a dream, but one toward which many scientists are working. Aquaculture requires control of the animals or plants in several ways other than control of the harvest as in the management of wild populations. The primary control over the reproduction, numbers, and varieties of the animals or plants usually requires that they be confined. Once this control is established, the control of nutrition and disease can follow. Aquaculture is conducted either by public agency or by private enterprise; if by the latter, it is usually operated in close association with agriculture.

The ancient art of aquaculture has had little scientific attention compared to agriculture. Only in recent decades have scientists begun intensive studies of such topics as fish genetics, artificial breeding of mollusks, nutritional requirements, pathology, or disease control. Many scientists contribute: the fishery scientist may be concerned with general cultural and breeding methods; the biochemist with nutritional needs for carbohydrates, fats or proteins; the pathologist and bacteriologist with disease identification and etiology. The need for better knowledge is great and the support for scientific study is rapidly growing.

1.7 Organization of This Book

A scientific tradition of long standing has divided the aquatic sciences into marine and freshwater parts. *Oceanography* is the scientific study of the physical, chemical, and biological features of the oceans; *limnology*, the parallel study of fresh waters. *Marine biology* and *freshwater biology* are commonly pursued in separate laboratories, although both are usually concerned with structure, physiology, behavior, and ecology of shallow-water aquatic animals and plants. The faunas and floras are mostly different because few forms can make the major physiological adjustment from fresh water to salt water, or vice versa, but nevertheless the fundamental scientific principles are the same.

The scientific studies of fisheries have been commonly divided also into freshwater and marine parts, but the practice has even less justification. The culture of carp has principles in common with the culture of oysters. The management of wild clam populations on a beach has principles in common with the management of Antarctic whales or lake trout in a large lake. A more natural division with respect to the resources is to be found between wild resources over which the only control is the amount of fishing and cultured resources in which control may be exercised also over migration, reproduction, nutrition, and disease. Such a division is strengthened because almost all of the wild resources are in public waters and almost all of the cultured resources are in private ownership (with the important exception of the fish culture that supports recreational fishing in several countries).

Still another division should be made between food and recreational fisheries, not because of differential biological principles of management, but because of their greatly different function in our societies. The former is using a resource that supplies a vital food and that should be economically and technically as efficient as possible, most people will agree. The latter is also filling an important human need, but one in which the concept of economic or technical efficiency in the catching of fish is completely irrelevant. The people who fish for fun frequently handicap themselves by light tackle and inefficient lures to the point of barely being able to catch a fish—all in the fulfillment of a sport. Their objective is recreation; the amount of food obtained is incidental.

Therefore, a discussion of the common scientific principles has been presented in Chapters 2–7. This includes merely a brief introduction to many of the sciences contributing to fisheries although in each case the matters relevant to the fishery problems have been emphasized. Much of this section is devoted to a fuller treatment of the analysis of fish populations, a topic which many consider to be the essence of fishery science. In many places, the fishery sciences are depicted as a rapidly developing body of knowledge, sciences in which major problems are being framed more rapidly than answers are being found, activities in which scientists can expect major challenges and the satisfactions of scientific discovery as well as of public service.

The numerous sciences contribute to knowledge of aquatic producing systems. These are of course ecosystems which man must manage if he is to gain maximum sustained benefit from them. The ecosystem concept forms a logical basis for the development of this book applying as it does to the broad principles of the human use of living resources as well as to details of the life of a fish. The concept unites the knowledge of the environment, the organisms, and the populations and their management in a multidisciplinary set of sciences.

After the introduction to the scientific principles, in Chapters 8 and 9 the recent history of food and recreational fisheries, the practice of aquaculture, and the general public strategy of managing the living aquatic resources are described.

The treatment of references is not in accord with usual scientific practice for two reasons. First, many of the general findings or concepts cannot easily be attributed to any one person because they have evolved through collaboration of many people, either by successive papers or through committee work. Second, the student should find comments on the particular value of major reference work more useful than a long list of references.

REFERENCES

One of the lucid projections of human population and resource problems and one that anticipated by 10 years the development of hard evidence about food shortage is:

Brown, H. (1954). "The Challenge of Man's Future." Viking Press, New York.

Authoritative recent summaries of the world's food problems are:

Food and Agriculture Organization of the United Nations (FAO). (1967). "The State of Food and Agriculture, 1967." FAO, Rome. See especially the review section on fisheries in the 1967 issue, pp. 119–144. This is one of an annual series, so look for the latest volume.
President's Science Advisory Committee (USA). (1967). "World Food Problem. Report of the Panel on the World Food Supply," Vols. 1 and 2.

For perceptive discussion of the food fishery problems in the middle 1960's, see the invited speeches by C. R. Lucas and W. M. Chapman to the Committee on Fisheries (COFI) of FAO on June 13, 1966 and April 24, 1967, respectively.

Earlier statements of the challenges as seen by two pioneering fishery scientists are:

Graham, M. (1943). "The Fish Gate." Faber & Faber, London.
Russell, E. S. (1942). "The Overfishing Problem." Cambridge Univ. Press, London and New York.

The Aquatic Environment

Water is an alien environment for us, very different from the land and surrounding air with which we are familiar. Aquatic animals and plants respire, feed, grow, escape enemies, and reproduce under conditions that we cannot experience and only vaguely understand. Yet almost every part of the aquatic environment is inhabited by species that perform all of these functions. They exist in wondrous varieties, some in enormous numbers, and others so rare that they are known only from single specimens. Just a few are useful to man, and these occur in abundance only in a small fraction of the waters. These few plants and animals comprise the organic production systems that are of concern to the fishery scientists. He must understand the functioning of the systems and the factors that restrict or enhance their production.

The production systems are complicated and incompletely understood. They are part of the concern of the marine and freshwater sciences that are collectively called *oceanography* and *limnology*, respectively. These are relatively new sciences, and they are rapidly adding to our knowledge of conditions and life in the waters. They include many matters of immediate importance to the fishery scientist, such as the productivity of the sea or the population dynamics of fish, as well as matters of remote interest, such as the geology of the sea bed or the fascinating biology of the coral reefs.

The oceanographers and limnologists, as well as geologists, geographers, engineers, and other applied scientists, are concerned also with water itself as a major resource. All of the basic physical and chemical studies of the natural waters help the applied scientists to understand and plan for the proper use of water for navigation, power, cooling, waste dilution, irrigation, and domestic supply. Similarly, the basic studies help with the solution of problems created when water causes damage through floods, droughts, or beach erosion.

Water for raising fish is just one of its many uses. These uses frequently conflict, but seldom is any one favored entirely because multiple uses are generally possible and desirable. When fishing is involved, the fishery scientist can help by seeing that the water is most effectively used for fish. He can do this best if he has an understanding of water as a resource as well as an environment producing fish.

It is beyond the scope of this volume to include a review of oceanography and limnology, but certain physical, chemical, and biological features of the waters which profoundly affect the production systems will be discussed. These are the features corresponding to the characteristics of the land that determine agricultural production, such as light, oxygen, climate, soil, plant nutrients, land topography, pests, and competitors. Special mention will be made of those aquatic features that are limiting or enhancing.

The summary to follow is necessarily descriptive rather than being theoretical or analytical. Moreover, all of the topics include difficult questions of physics, chemistry, and geology that have received intensive research. The treatises listed under References will provide an introduction to the extensive literature in each field.

2.1 Hydrology and the Water Cycle

Hydrology is broadly defined as the science dealing with the occurrence, circulation, distribution, and properties of the waters of the earth and its atmosphere. It embraces many sciences, including oceanography, limnology, and meteorology, but its usage is becoming restricted to the occurrence and distribution of water over the land. Hydrologists are commonly concerned with the development and use of water supplies on and under the surface of the land.

Water covers about 71% of the earth's surface. More than 98% of this accessible water is in the oceans, and most of the balance is in the accumulation of ice on the land masses. Only about 0.02% is fresh and occurs in atmospheric water vapor, inland waters, and circulating groundwater. This fresh water is estimated to be somewhat less than the amount that evaporates annually from the oceans and the land (Table 2.1).

Despite the relatively small proportion of the earth's water that is transferred each year from the oceans to the atmosphere and the land, this part is extremely important. It forms the hydrologic or water cycle (Fig. 2.1). About 1 m of water evaporates annually from the surface of the oceans

TABLE 2.1
Area and Volume of Water on Earth[a]

Area of earth's surface	510	$\times 10^6$ km^2
Area of all land	149	$\times 10^6$ km^2
Area of all oceans and seas	361	$\times 10^6$ km^2
Area of Atlantic Ocean	82	$\times 10^6$ km^2
Area of Pacific Ocean	165	$\times 10^6$ km^2
Area of Indian Ocean	73	$\times 10^6$ km^2
Volume of all oceans and seas	1370	$\times 10^6$ km^3
Volume of polar caps and other land ice	16.7	$\times 10^6$ km^3
Volume of inland waters	0.025	$\times 10^6$ km^3
Volume of circulating ground water	0.25	$\times 10^6$ km^3
Volume of atmospheric water vapor	0.013	$\times 10^6$ km^3

[a]Source: Sverdrup, Johnson, and Fleming (1942) and Hutchinson (1957).

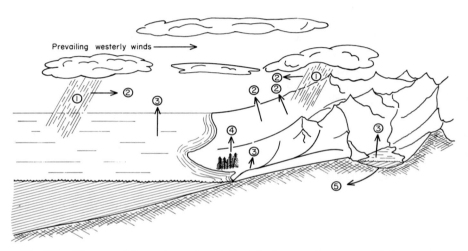

Fig. 2.1. Diagram of the water cycle. (1) Precipitation as rain or snow. (2) Evaporation from falling particles and from the earth. (3) Evaporation from bodies of water. (4) Transpiration from plants and animals. (5) Drainage underground.

TABLE 2.2
Approximate Water Balance of Oceans and Continents[a]

	g/cm^2/year	km^3/year $\times 10^3$
Evaporation from ocean surfaces	106	383
Precipitation on ocean surfaces	96	347
Evaporation from land surfaces	42	63
Precipitation on land surfaces	67	99

[a]Source: Hutchinson (1957). These values are not known accurately and other figures will be found.

(Table 2.2) of which about 91% returns to the ocean in the form of rain, the remaining 9% returns to the land. Of the latter, about one-fifth appears in the lakes and rivers, about one-fifth enters the soil as groundwater, and about three-fifths evaporates from free water surfaces and from plants.

The average precipitation on land of 67 g/cm²/year is distributed very unevenly. More than the average falls from the westerly winds of the temperate zones, especially near the mountains of the western sides of the continents. Much less falls from the easterly trade winds of the subtropical areas, although in the doldrums near the equator, more than the average falls. Secondary monsoon systems develop in some subtropical parts of the world, where almost all of the precipitation falls in a part of the year. Even in the wetter temperate zones precipitation is usually greater in winter, evaporation is greater in summer, and some precipitation is stored temporarily as snow or ice. Such seasonal variability results in cycles of change in the flow of rivers or level of lakes; sometimes the changes are great enough to cause seasonal floods or drying of the flow.

The fishery scientist should always recognize that the fish must endure the extreme conditions which are commonly present in rivers and estuaries. The low waterflow may also be critical for irrigation, waste dilution, power, or other uses and makes it even more difficult to accommodate the needs of the fish.

2.2 Bottoms and Basins

Anyone who has watched a muddy river flood a valley or noticed the changes in a beach during a storm must appreciate the role of water as a transporter of sediments. The river bottom quite literally flows toward the sea as the water carries the finer particles and rolls the bigger ones. The waves pound the beach to push some particles higher and carry others back out, and soon a bar appears. In addition the wind blows dusts, and ice carries rocks and other debris far out over the water. Thus the bottom underneath the water is covered with combinations of clay, silt, sand, gravel, stones, and boulders. To these are added the skeletons and wastes from plants and animals, and particles blown out of volcanos or formed by direct precipitation of chemicals from the water. The result is an infinite variety of bottom sediments from the finest clays consisting of particles averaging less than 0.001 mm in diameter to boulders (Table 2.3).

Sediments cover almost all of the bottom, from the shores to the most distant ocean deeps. Bed rock is a rarity found seldom except in the cliffs of mountains or canyons which may mark the shore or be found far beneath the surface. The sediments in lakes and near the ocean shores, called

TABLE 2.3
ATTERBERG'S SIZE CLASSIFICATION OF
SOIL PARTICLES[a]

Grade limits (mm)	Name
2000–200	Blocks
200–20	Cobbles
20–2	Pebbles
2–0.2	Coarse sand
0.2–0.02	Fine sand
0.02–0.002	Silt
< 0.002	Clay

[a]Adopted by the International Commission on Soil Science.

terrigeneous, are generally coarse and mostly sand. The inorganic fraction of sediments of the ocean depths far from shore are generally fine. Such *pelagic* deposits may be largely of inorganic origin in which case they are known as red or brown clay; if they contain significant amounts of skeletal materials they are called *oozes*.

The study of the transportation of sedimentary particles and the formation of sediments is largely in the field of geology. Geologists are discovering many clues to the history of the earth beneath the water, partially because most of the rock now exposed at the surface of the earth was once a sedimentary deposit at the bottom of the sea. Deposits containing large amounts of organic matter may have produced petroleum.

The geologist's findings about the existing sediments are of interest to fishery scientists because a large fraction of the useful aquatic plants and animals live close to the bottom. Here they may attach, burrow, rest, hide, place their eggs, or feed on bottom-living plants and animals. Each type of bottom will favor different plants and animals according to their particular adaptation. A large part of the world's fisheries depend on species collectively called the *bottom fishes*—fishes that are caught within about 5 m of the bottom.

2.2.1 LAKES AND RESERVOIRS

Water accumulates in a great variety of basins that have been formed by tectonic, volcanic, glacial, fluviatile, and other natural processes. In addition, man constructs reservoirs to store water where and when he needs it. Most lakes are parts of river systems, but some are in closed basins from which water escapes only by underground flow or by evaporation. The largest body of water in a closed basin, the Caspian Sea, has an area of 436,000 km². The largest lake in an open basin, Lake Superior, is 83,000 km². The deepest known, Lake Baikal in Siberia, is 1741 m.

The form of the basins is shown on *bathymetric* charts which have contours (isobaths) of equal depths. From such charts are obtained the essential descriptive measurements or data for computation of length (l), breadth(b_x), maximum depth (z_m), area (A), volume (V), mean depth ($\bar{z}$ or V/A), and the length of shore line (L).

From these data are calculated two measurements that are useful in a comparison of different lakes with respect to their productivity. One of these is the index of shoreline development (D_L), a ratio of the length of shoreline to the circumference of a circle with an area equal to that of the lake [$D_L = L/2(\pi A)^{\frac{1}{2}}$]. A D_L of 1 is a perfect circle; a larger index indicates the departure from a circle. The other is the index to the shape of the basin, a ratio of the mean depth to maximum depth ($\bar{z}/z_m$). This index has a value of 1.0 in a cylinder, 0.33 in a perfect cone.

Lakes have been classified according to the form of their basins in too many ways to repeat here. One classification in wide use makes a division into two types on the basis of the relative amount of life, a division closely related to the kind of basin. *Oligotrophic* lakes have scarce populations of plants and animals. They are usually deep, cold, and clear. Sparce sediments occur on the bottom, and oxygen is present at all times in deep water. They tend to have a low index of shoreline development (D_L) and a large ratio of mean depth to maximum depth ($\bar{z}/z_m$). *Eutrophic* lakes have abundant populations of plants and animals and organically rich bottom materials. Usually they are relatively shallow, warm, and turbid, with low oxygen at times in the deeper water layers. They tend to have a high index of shoreline development and a small ratio of mean depth to maximum depth.

2.2.2 ESTUARIES AND LAGOONS

Near the shores of the oceans are many bodies of water either continuously, intermittently, or not connected with the ocean. They have been classified into two major groups, not on the basis of the type of basin which may be highly diverse, but on the basis of the circulation of their water. *Estuaries* are where fresh water and salt water meet and mix. The most generally accepted definition is a semienclosed body of water that has a free connection with the open sea and within which seawater is measurably diluted with fresh water derived from land drainage. Excepted from this definition are the large semienclosed seas, such as the Baltic. *Lagoons* are coastal bodies of either fresh water or salt water that may have an intermittent connection with the ocean but which usually have a stable salinity and little or no tidal exchange. They may be much saltier than the ocean if water flows occasionally in from the sea and evaporates. They may be fresh if elevated slightly above the ocean. They are usually small and shallow and can be considered as another variety of lake.

Estuaries vary as environments according to the mixing process which is determined largely by the tides, by the inflow of fresh water, and by the shape of the basin. The extremes are the shallow estuary that has an equal salinity from top to bottom and the deep estuary into which both river and sea flow and out of which flows a mixture of fresh water and salt water on the surface (see Section 2.6.5). The classification of estuaries has therefore been developed with consideration of both geologic origin and the mixing processes which occur only partly as a consequence of the shape of the basin. No standard classification exists, but the following are the major types of estuaries in approximate order from the shallowest to the deepest:

1. Deltaic river mouths, such as the Mississippi, Amazon, and Niger. In these the river flows almost directly into the sea and there is almost no mixing of fresh water and salt water. Salt water occurs only as a wedge along the bottom.
2. Bar-built estuaries, such as the inland seas of the Netherlands. In these large, shallow seas the tidal turbulence and winds mix the water almost uniformly from top to bottom in most parts.
3. Drowned river mouths, such as Chesapeake Bay. In these the fresh water and salt water are mixed considerably, but not completely, from top to bottom.
4. Tectonic estuaries, caused by faulting and submergence of the land, such as San Francisco Bay. These are usually deep. Mixing of fresh water and salt water occurs in the surface layer only.
5. Fjords caused primarily by glacial action such as those in Norway, Chile, and Alaska. These are usually very deep. Mixing of the water occurs only in the surface layer.

2.2.3 OCEANS

The immensity of the ocean basins is not entirely indicated by the proportion of the earth's surface in the oceans (71%) because much of the oceans is very deep. The average depth is 3800 m, about 4.5 times the average elevation of the land (840 m) above sea level. Great depths occur in the *trenches*, several of which are more than 10,000 m deep. The deepest point now known is in the Mariana Trench—11,033 m. Such depths are substantially more than the altitude of the highest mountain—Mt. Everest is 8848 m high.

Charting the ocean depths has been undertaken for many centuries along the coasts where hazards to navigation exist. Most maritime nations have published charts of their coastal waters. Usually these charts show *isobaths*, lines of equal depth, in areas where the soundings are numerous, but beyond a short distance from the coast the soundings themselves are shown frequently because few reliable soundings have been obtained in the depths of the oceans. Fortunately much more detailed charts have become available since the development of echo sounders about four decades ago. The deep-sea charts now show *ridges, rises, trenches, seamounts, canyons, basins,* and other features. Much more remains, however, to be discovered and charted.

The geology of the deep-sea floor is a vital part of the knowledge of the earth's crust, so it is surprising that most of the knowledge of the sea bottom, its sediments, and topography has come during the last four decades— indeed, largely since the end of World War II. It remains one of the least-known features of the earth, whose exploration is considered as challenging to many people as the exploration of outer space.

Fortunately for the fisheries the knowledge of the area above the 500-m depth, which is about the maximum that can be reached by ordinary commercial fishing gear, is much more precise. Most of it has been thoroughly charted, not only with respect to the depths and hazards to navigation, but also with respect to the fishing banks, bottom types, and hazards to fishing gear.

The predominant feature of the coastal zone, the *continental shelf*, is of utmost importance to the fisheries. It is the place where most of the world fishing occurs either with gear that catches the bottom fishes, or with gear that catches the pelagic, schooling fishes in the water above the shelf.

The continental shelf is indeed a shelf—a nearly flat plain extending out to a margin beyond which the bottom slopes steeply and is furrowed by canyons (Fig. 2.2). This zone is called the *continental slope*. The shelf margin occurs at an average depth of about 130 m, but most charts show an isobath at 100 fathoms (fm), or 200 m which for practical purposes marks the margin of the continental shelves of the world. The depth of 200 m has been designated in some international treaties as the margin of the continental shelf.

The exact depth at the margin is not especially important to the fisheries because seaward from the actual margin the continental slope is usually very steep, rough, and difficult to fish. Minor *terraces* of flat bottom occur in a few places on the continental slope, but these are seldom important to the fisheries. The proportion of the ocean floor less than 200 m deep, most of which is continental shelf, is estimated at 7.6%. The proportion between 200 m and 1000 m is only 4.3%. When the latter figure is interpolated, it may be estimated that the area between 200 and 500 m is only 1.6% of the ocean floor. The other 90.8% of the ocean floor is beyond the reach of most commercial fishing gear.

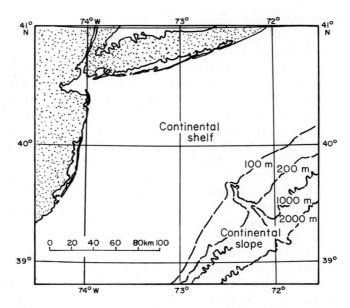

Fig. 2.2. Bottom topography of part of the continental shelf and slope off eastern United States.

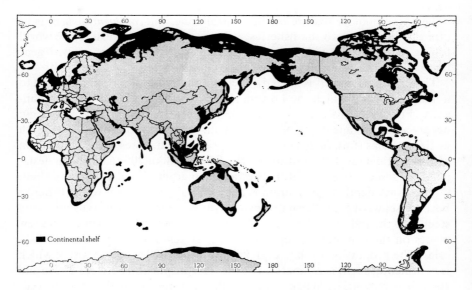

Fig. 2.3. The continental shelves of the world [Source: FAO Doc. 57/7/4725. A contribution to the UN Scientific Committee on the effects of atomic radiations on the specific questions concerned with the oceanography and marine biology in respect to the disposal of radioactive wastes. 82p.]

The continental shelves of the world are widest off the coastal plains of the land and narrowest off the land with mountain ranges. Their widths range from almost 0 to 1300 km, with an average of about 50 km. The shelf slopes an average of only about 0.2%, a slope so small that the shelf would appear level to the human eye. On the other hand, the average slope of the continental slope varies from about 3.5% off the flat coastal plains to about 6% off the mountainous coasts.

The major continental shelves are off northwest Europe, off Siberia in the Arctic Ocean, off Alaska in the eastern Bering Sea, in the several seas off eastern Asia, the Argentine Shelf off eastern South America, and off the North American coast between Cape Hatteras and Newfoundland (Fig. 2.3). All except the Siberian shelf are major fishing grounds.

On the continental shelves occur *banks* which are elevations frequently important to fishermen but which are not hazards to navigation, *shoals* and *bars* which are nonrocky hazards to navigation, and *reefs* which are coral or rocky hazards to navigation.

2.3 Physical Properties of Water

The fishery scientist needs to understand some of the physical properties of water that affect its function as an environment for plants and animals. Some of the properties may influence the behavior of the organism, e.g., temperature and light transmission; others may be useful as a means for fishermen to find animals, e.g., sound transmission; still others may be useful in understanding the circulation of the water.

2.3.1 TEMPERATURE

The temperature of the water influences profoundly the lives of all aquatic plants and animals. Plants and most of the animals assume the same or almost the same temperature as the water. Such animals are called *cold blooded*, or *poikilothermous*. Each plant or animal is adapted to a normal seasonal temperature regime and is commonly affected adversely by unusual temperatures. Many animals reproduce, feed, or migrate only within certain temperature limits which may be narrow or broad according to the species (Table 2.4, Fig. 2.4).

Temperature also controls the density of water in ways that determine the entire temperature structure of all waters. It changes the solubility and physiological effects of solids and gases so that their effect on animals must be considered together with temperature. For a fishery scientist temperature is commonly the most important item of physical information about a body of water.

TABLE 2.4

ᴀᴘᴘʀᴏxɪᴍᴀᴛᴇ ʀᴀɴɢᴇ ᴏꜰ ᴛᴇᴍᴘᴇʀᴀᴛᴜʀᴇs ᴛᴏʟᴇʀᴀᴛᴇᴅ ʙy
Sᴏᴍᴇ ʀᴇsᴏᴜʀᴄᴇ ᴀɴɪᴍᴀʟs

Yellowfin tuna	18°–32°C
Cod	1.0°–12°C
Pompano	15°–35°C
Rainbow trout	0°–20°C
American oyster	1°–36°C
(on exposed tide flats)	–49°C
Pacific salmon	
Fresh water	0°–27°C
Ocean	0°–20°C

The earth and its atmosphere as a whole are warmed by radiation from the sun and cooled by an equal amount of radiation back to space; thus on the average the temperature remains the same. The warming and cooling by radiation is not uniform in different parts of the earth's surface; the low latitudes are warmed more than they are cooled and the higher latitudes are cooled more than they are warmed. Both ocean and land transfer heat to the atmosphere. Wind and ocean currents move the heat around, generally from

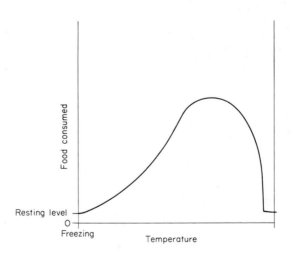

Fig. 2.4. Diagram of relation of rate of food consumption of cold-blooded animals to temperature. [After N. S. Baldwin (1957). Food consumption and growth of brook trout at different temperatures. *Trans. Amer. Fish. Soc.* **86**, 323–328.]

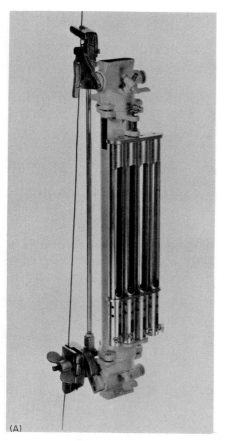
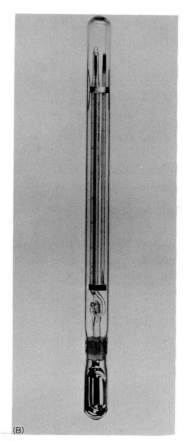

(A)   (B)

Fig. 2.5. A Nansen Bottle (A) and a reversing thermometer (B). The bottle is lowered with the valves open. When it is at the desired depth, a weight is dropped down the wire to strike the lever at the top and allow the bottle to topple over closing the valves and inverting the thermometers. Each thermometer contains two mercury columns, in one of which the column breaks when inverted to record the temperature and the other of which is used to correct the reading for the ambient temperature when read at the surface. (Photos courtesy of G. M. Mfg. and Instrument Corp., New York.)

the tropics toward the poles. The balance of the heat received, radiated back, and transported latitudinally also remains about the same; thus the mean annual temperatures of both air and water at each point on the earth remain about the same.

The waters are heated almost entirely by radiation from the sun. The radiation is almost all absorbed, and heat is lost continuously by back radiation, evaporation, and conduction. The waters are warmed by day and cooled at night, gradually warming more during the spring and summer and cooling during the autumn and winter. In temperate regions the waters are usually warmest in late summer, coldest in late winter.

The radiation from the sun warms the waters only in a thin surface layer. More than 90% of the heat is absorbed in the upper 20 m of the clearest ocean waters and in the upper 4 m of the average coastal waters. All waters below are warmed by mixing with the surface layers.

Temperatures are measured with a variety of instruments which are chosen according to the depth, accuracy, and frequency of measurements desired. An occasional surface temperature is measured easily by a conventional mercury thermometer, but if a continuous record is desired, an electrical resistance thermometer may be installed with a recorder. At depths below the surface that can be reached with instruments on a wire special reversing thermometers are used (Fig. 2.5). These are designed to indicate the temperature in a conventional manner when in one position but retain the same reading indefinitely after being inverted. They are lowered in the recording position and inverted by a "messenger" dropped down the wire. These thermometers are made to measure temperatures within $\pm 0.02°$C, an accuracy needed for subsequent computation of density (see Section 2.3.3). In shallow water (< 100 m), electrical resistance thermometers are also used when accuracy need be no closer than $\pm 0.5°$C.

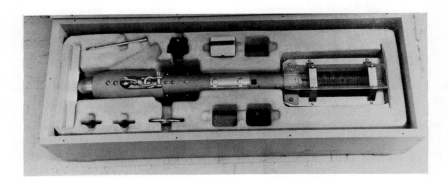

Fig. 2.6. A Spilhaus bathythermograph in its storage case. The instrument contains a pressure element and a temperature element. It produces a graph of temperature vs. depth to a maximum depth of 300 m. (Photo courtesy of G. M. Mfg. and Instrument Corp., New York.)

One of the most convenient instruments for repetitive measurements of temperature in the upper 500 m of water is the bathythermograph (Fig. 2.6). This is a rugged instrument that may be lowered to depth from a moving vessel. It contains a pressure sensor and a temperature sensor which record a graph of temperature against depth on a glass slide.

The maximum surface temperatures in fresh water and salt water (excluding hot springs) occur in shallow sheltered waters where 40°C may be attained. Temperatures at the surface of the open sea and of large lakes rarely exceed 30°C. The minima are the freezing temperatures 0°C in fresh water and about − 1.9°C in seawater. The seasonal variation is related to the size of the waters and is greater close to land. There, the water may reach 40°C in the heat of a summer day and may freeze in winter. Open ocean water shows commonly a change in temperature of less than 4°C in the tropics and the polar regions and less than 10°C in the intervening temperate zones (Fig. 2.7).

The surface temperatures prevail only in the upper layer which is mixed by the wind. At the bottom of the mixed layer is a *thermocline*, a layer in which temperature changes rapidly with depth (see Section 2.3.3). Below the thermocline (except in tropical lakes) temperatures range from about 4°C in lakes down to about 1°C in the great depths of the oceans. Thus ocean and lake waters below the thermocline are almost constantly and nearly uniformly cold.

2.3.2 PRESSURE AND COMPRESSIBILITY

The pressure at any depth in the water is equal to the weight of the atmosphere and the weight of the water column. For rough calculations the latter may be estimated as the equivalent of 1 atm for each 10-m increase in depth. One atm is equal to the weight of 760 mm of mercury; about 14.7 pounds/inch^2, or 1.013 kg/cm^2.

Water is only slightly compressible; for biological purposes it may be considered as not changeable in volume at all. Much more important biologically is the differential compressibility of water and gases. The latter compress proportionately to pressure. Any gas in the lungs or swim bladder of an animal swimming from the surface to a depth of only 10 m will compress to half of its volume at the surface. It will compress to only one-tenth of the volume at the surface when the animal goes to 90 m. When gases do not occur in animals and plants they are little affected by changes in pressure of 10 or 20 atm but may be seriously affected by changes of 200 atm. Such effects probably arise from the slightly different compressibility of water and skeletal parts.

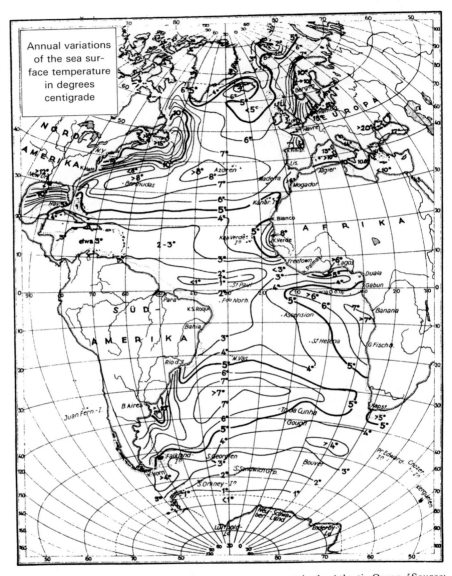

Fig. 2.7. Annual variations of the sea surface temperature in the Atlantic Ocean. [Source: G. Schott (1942). "Geographie des Atlantischen Ozeans." Boysen, Hamburg.]

2.3.3 DENSITY

The density of water is controlled by temperature, dissolved solids, and pressure. It is usually defined in relation to the density of pure water at its temperature of maximum density which weighs 1 g/cm^3. The relative density of water at different temperatures, salinities, and pressures is shown in Tables 2.5 and 2.6.

The density of seawater is generally between 1.01500 and 1.03000, but oceanographers rarely report it in this way. Instead they use a convention of

TABLE 2.5

DENSITY OF WATER AS A FUNCTION OF TEMPERATURE AND SALINITY[a]

Salinity (%o)	Temperature (°C)				
	0	5	10	20	30
0	0.9999[b]	1.0000	0.9997	0.9992	0.9957
15	1.0120	1.0119	1.0114	1.0096	1.0069
25	1.0201	1.0198	1.0192	1.0172	1.0143
35	1.0281	1.0277	1.0270	1.0248	1.0218

[a]Source: M. Knudsen, ed. (1901). "Hydrographical Tables." G.E.C. Gad, Copenhagen.
[b]Density of ice at 0° C and 0%o salinity is 0.9168.

TABLE 2.6

DENSITY OF PURE WATER AS A FUNCTION OF DEPTH AT 0°C[a]

Depth (m)	Density
0	0.9999
250	1.0009
500	1.0020
1000	1.0042
2000	1.0084

[a]For more complete tables, see M. Knudsen, ed. (1901). "Hydrographical Tables," G.E.C. Gad, Copenhagen.

subtracting 1.00000, multiplying by 1000, and designate the value as σ. Density can be measured directly by simple hydrometers when high accuracy is not required as in estuaries where σ may range from 0 to 25. Such determinations can be reliable to about $\pm0.2\sigma$, but this accuracy is not enough for work in lakes or oceans. In such cases density is usually calculated from temperature and salinity, both of which can be measured very precisely (Fig. 2.8).

Changes in density are so slight that they have little if any direct biological significance that can be separated from the individual effects of temperature, salinity, and pressure; but they are of great significance for the indirect estimation of water currents (see Section 2.6.3).

The vertical changes in temperature and the associated changes in density are, however, of great biological significance. Water near the surface is most dense at a temperature of about 4°C when fresh or between 4 and −2°C when salt. The water of lowest density is found at the surface. When density increases with depth are large, the water column is resistant to mixing by wind and current; when density differences are very small, the water column can be mixed to great depths. In consequence the seasonal changes in surface temperatures are accompanied by changes in the temperature structure of the water column.

Consider a lake in a temperate region in early spring when the entire water column is near the temperature of maximum density (4°C). A condition of instability prevails when the entire lake can be stirred by the wind (Fig. 2.9).

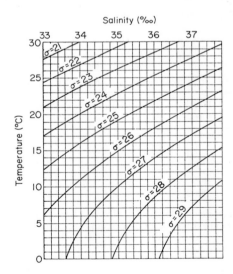

Fig. 2.8.   Temperature–salinity–density relationship.

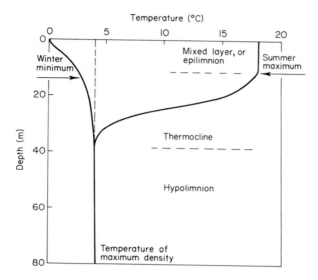

Fig. 2.9. Diagram of water temperature in a lake at different seasons. As the surface temperature warms from the winter low, the water column becomes unstable at 4°C and then stable again as it warms further. In autumn, the reverse occurs.

As the surface water warms with the season, the mixing cannot distribute the warmed water evenly; and because the warmer water is less dense than the cooler water it remains on the surface. This layer will be mixed by wave action to a depth of only a few meters. Below this mixed layer will be a layer with a large temperature gradient, called the *thermocline*. Below the thermocline the water will remain at the temperature it had when last overturned, near 4°C, the temperature of maximum density in deep temperate-zone lakes. In the ocean the process and temperature structures are similar (Fig. 2.10), but the density is modified also by the salt content, the mixing processes are more complicated, and the temperature at maximum density is below 0°C in typical seawater.

The thermocline disappears seasonally in most freshwater lakes. When the surface waters cool in autumn, the density increases are again unstable. If the annual cycle of temperature change goes below and above the temperature of greatest density, two periods of mixing occur. In arctic lakes where the temperature increases merely to 4°C or less, only one period of mixing will occur. Likewise, in tropical lakes only one period of mixing will occur at the time of greatest cooling; the temperature below the thermocline may be much above 4°C. In the case of the oceans the thermocline is usually permanent in tropical seas, seasonal in temperate waters, and absent or poorly developed in arctic waters. In a few lakes and isolated ocean basins of

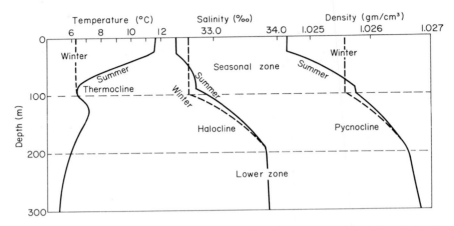

Fig. 2.10. Diagram of temperature, salinity, and density structure of the Subarctic Pacific Ocean off British Columbia [Source: J. P. Tully (1965). Time series in oceanography. *Trans. Roy. Soc. Can.* **3**(4), 366.]

extreme salinity unusual temperature conditions prevail—the discussion of which is beyond the scope of this volume.

The thermocline is a barrier, a zone of stable water through which mixing of the waters does not occur. It is a zone of large temperature change which must be accommodated by animals swimming through it. In the ocean it may be at a depth of 1000 m or as shallow as 10 m; in lakes it is usually found between 5 and 50 m. In places sheltered from the wind its depths may be

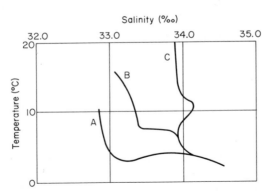

Fig. 2.11. Temperature–salinity relations of water masses of northeast Pacific Ocean. (A) Subarctic water in gyre of central Gulf of Alaska. (B) West wind drift near 45°N lat. (C) Subtropical water near 40°N lat. [Adapted from S. Tabata (1965). Variability of oceanographic conditions at Ocean Station "P" in the northeast Pacific Ocean. *Trans. Roy. Soc. Can.* **3**(4), 367–418.]

constant; in other places its depth may change rapidly as winds and currents pile up or disperse the mixed surface layer.

Because density depends upon both salinity and temperature different combinations of salinity and temperature can produce the same density. Consequently the salinity–temperature–depth relationship is used directly to identify water masses. When temperature and salinity of a column of water are plotted, the shape of the curve is frequently characteristic of water from a particular region (Fig. 2.11).

2.3.4 SOUND

Poets who have written emotionally about the quiet waters have been unaware that there are numerous sounds in water. The natural sounds of waves, cracking ice, and animals are transmitted in water much more efficiently than sounds in air. They do not pass efficiently through the air–water surface so we do not hear the sounds from below.

Sounds are produced by a large proportion of the aquatic mammals, fish, and shrimp. They produce sounds, either by forcing air from lungs or air bladders, over vibrating membranes, by grinding teeth, by rubbing fins, or by snapping together parts of their skeletons. They also produce mechanical sounds incidentally when either swimming, burrowing, or feeding, the latter especially when eating crustaceans or mollusks. Their sounds have been described as thumps, grunts, growls, groans, knocks, thuds, clucks, boops, barks, whines, grates, scrapes, clicks, chirps, and snaps. The sounds may occur in a great chorus, as from a school of snapping shrimp. Most of the sounds produced by animals are of low frequency varying from about 10 to 1000 Hz (cycles/second), but some produced by mammals are of frequencies ranging up to 80,000 Hz, far above the limit of human hearing (about 10,000 Hz).

Sound is important to the fisherman who uses either echo-sounding or echo-ranging equipment for either navigation or fish location. (*Echo sounding* is used here to designate the practice of directing a sound beam vertically down to the bottom and receiving an echo; *echo ranging*, the practice of directing a beam horizontally or at a substantial angle from the vertical.) Echo-sounding equipment is considered essential for all high-seas vessels as a navigational aid and as a help in locating fish near the bottom (Fig. 2.12). Echo-ranging equipment is used extensively by vessels for the location of fish in the depths midway between surface and bottom. Some wholly new fishing techniques have developed around the use of these sonic devices.

Sound in water travels between 1400 and 1550 m/second. It travels slowest at the surface of fresh water at 0° C and increases in speed with increasing

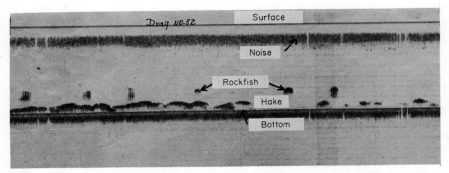

Fig. 2.12. An echo-sounder record of schools of hake and rockfish in 150 m of water in Puget Sound. (Record by W. Pereyra, U.S. National Marine Fisheries Service.)

temperature, salinity, and pressure. The speed near the surface is about 4.0–4.6 times as fast as the speed in air.

Sound is reflected when it strikes the surface of material of different properties. It is reflected strongly by an air–water surface and by a rocky bottom, less strongly by a mud bottom and by the skeletons of animals, least by the water-filled, fleshy bodies of plants or animals. It is reflected strongly by the air bladders of fish. Modern instruments will produce echo records that make it possible to distinguish between different types of bottom and various kinds of fish.

Sound not only is reflected by the surface and the bottom but also it is refracted (bent) by waters transmitting sound at different speeds, especially in the region of the thermocline. Consequently echo ranging at angles other than vertical frequently requires interpretation of multiple echoes that may be caused by reflection, refraction, or both. Such difficulties severely restrict the usefulness of near-horizontal echo ranging for fish location to 1000 m or less in many waters.

2.3.5 ELECTRICITY

Water conducts electricity with far less resistance than air. Specific resistance is that offered by a 1 cm cube with the current perpendicular to two parallel faces. The specific resistance, expressed in ohms, is the reciprocal of the specific conductance, a measure which is convenient to use at times.

The specific resistance of water varies according to salinity and temperature. The least saline natural waters have a specific resistance of about 50,000 ohms; average fresh water, about 6000; and hard waters, about 3000 ohms. Average seawater of 35‰ (per mille) salinity has a resistance of only about 18 ohms at 25°C, 33 ohms at 0°C.

Such a conductor as seawater moving through the earth's magnetic field produces a gradient of electric potential which is strong enough to measure with field instruments. The electric potential is roughly proportional to the velocity of the water current, being altered somewhat by the conductance of motionless water layers or the bottom and by the imposition of earth currents. The natural electrical potential near the surface of the ocean varies from less than 0.05 μV/cm to about 0.05 μV/cm. The lesser value is close to the minimum practical limit of field instrumentation, so the (presumably) smaller potentials in fresh water have not been measured similarly. (See Section 2.6.3.)

2.3.6 LIGHT

In addition to warming the surface of the waters the light of the sun provides the energy for photosynthesis of organic material by plants and enables all of the animals that have eyes to see. It varies in daily and seasonal cycles, and these cycles regulate the metabolism and behavior of most animals and plants. The direction of the sun may be used by some animals to guide their migrations. The sun has a vital influence on all life; indeed it sustains life itself.

Natural waters reflect, scatter, and absorb light so greatly that only a thin surface layer is ever illuminated. In the clearest ocean water, only 1% remains at 150 m; in average coastal water, at 10 m; in really turbid water, at only 1 m (Fig. 2.13). Men who descend in vehicles can barely see light from the sun below about 700 m in the clearest ocean water, and only 80 m in

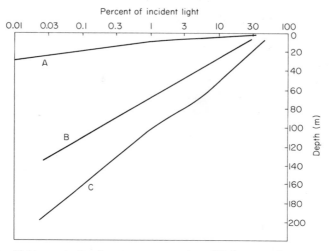

Fig. 2.13. Light penetration in some ocean waters. (A) Average of some coastal waters. (B) Caribbean Sea. (C) Sargasso Sea, one of the clearest of ocean waters.

average coastal water. Some fish may have more sensitive eyes than man, but we have no way of knowing exactly what they can see.

Transmission of the different colors depends on the dissolved pigments, on optical properties of pure water, and on suspended materials in the water. In the clearest lake and ocean water blue light is transmitted best, hence the deep blue appearance of many large lakes and seas. The transmission of blue light is affected more than other colors as the amounts of dissolved pigments and suspended material (including one-celled plants) increases. Coastal water appears green, and turbid water appears yellow-brown. All waters are relatively opaque to red light.

Transparency of surface waters is commonly measured by a Secchi (pronounced sek-key) disc 20 or 30 cm in diameter (Fig. 2.14). It is lowered on a graduated line until it just disappears and then raised until it just appears; the average of the two depths is called the Secchi disc transparency. This measure of transparency ranges from about 60 m in the clearest ocean water to a few centimeters in the turbid water of river mouths. It is equivalent to the level of penetration of between about 10 and 15% of solar radiation. More precise observations and measurements of waters other than the surface layer require optical equipment, but such measurements are rarely made during fishery studies.

The light of the sun is not the only light in the water; many animals and plants produce their own. They are the luminescent forms which are abundant in the sea but rare on land and in fresh water. Many bacteria, phytoplankton, jellyfishes, worms, crustaceans, mollusks, and fish produce light. They can be seen at night in most warm seas when the water is disturbed by either a wave, a propeller, or an oar. Many of them live in a twilight zone moving up at night and down at daylight. Many of the fish that produce light also have large eyes which may enable them to see at unusually low levels of illumination.

The ecological significance of the bioluminescence is not well understood. It is speculated that light flashes may confuse predators, attract prey, or serve to communicate with others, but we really do not know.

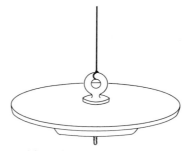

Fig. 2.14. A Secchi disc used for estimation of the transparency of the surface water layer.

2.4 Dissolved Materials

All aquatic organisms maintain a constant exchange of dissolved materials between their bodies and the surrounding water. Some materials such as dissolved oxygen and carbon dioxide are indispensable for life itself. Other materials such as the nutrient materials determine the productivity of the water, and the salts of the sea influence greatly the density of the water and pose a condition to which all organisms must adapt.

2.4.1 OXYGEN

Next to temperature the dissolved oxygen content of the water is probably the most common measurement of biological significance. Oxygen is essential to almost all life so its shortage or absence commonly limits the distribution of plants and animals. Some fish in warm water may be distressed at concentrations less than 5 mg/liter, others at lower temperatures may tolerate less than 1 mg/liter. Unlike the oxygen of the atmosphere which occurs in a nearly constant proportion of 20.9% the oxygen dissolved in water varies from none to twice or more the saturation value of about 0.7% by volume or 0.0001 % by weight, or as it is commonly expressed, 10 mg/liter.

The amount of oxygen that dissolves in water from the air depends on barometric pressure, temperature, salinity, and proportion of oxygen in the air. The pressure at the surface of the water near sea level is usually close to one standard atmosphere, and the proportion of oxygen varies but slightly depending mostly on the amount of water vapor in the air, so these factors have only a slight influence. Temperature and salinity, however, cause large changes in the saturation levels (Fig. 2.15).

The quantity of dissolved oxygen in water has been determined traditionally by the Winkler titrimetric method, but recently electrical methods for continuous measurement have been developed. Descriptions and instructions will be found in standard reference works.

Varied and sometimes confusing methods of reporting the quantity of oxygen are in common use, including proportion by weight, proportion by volume, and percent saturation (Table 2.7). The proportion by weight is usually used for fresh water and is expressed as parts per million (ppm) or milligrams per liter, with the liter at 4° C, the temperature at greatest density. Oxygen in seawater will be found expressed as milligram atoms per liter (mg-atoms/liter), or as milliliters per liter, with the liter at 20° C. In addition, the percent saturation is used for both seawater and fresh water because temperature and salinity have large effects on solubility. Percentage saturation is useful because it suggests the recent biological history of the water. It is computed as 100 times the quantity observed divided by the estimated saturation value for the temperature and salinity of the sample *in situ* but

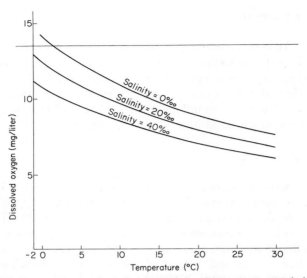

Fig. 2.15. The solubility of oxygen in equilibrium with water-saturated air at a pressure of 760 mm Hg. [Source: G. A. Gruesdale, A. L. Downing, and G. F. Lowden (1955). The solubility of oxygen in pure water and sea water. *J. Appl. Chem.* **5**, 53–63.]

at surface pressure. At sea level the standard atmospheric pressure of 760 mm Hg is used, but for inland waters it is corrected for altitude. The saturation at depth is not reported with consideration of the increased solubility due to water pressure.

Most oxygen in natural waters comes from the air by solution at the water surface; some comes from photosynthesis in the illuminated layer. Oxygen is removed primarily by respiration of plants, animals, and bacteria at all depths; secondarily by transfer from supersaturated surface waters back to the atmosphere; and thirdly by chemical reactions.

It would seem reasonable to assume as the early scientists did that the

TABLE 2.7
APPROXIMATE MULTIPLICATION FACTORS TO CONVERT UNITS OF OXYGEN
CONCENTRATION

To:	ppm, mg/liter	ml/liter	mg-atoms/liter
From: ppm, mg/liter	1.0	0.70	0.0625
ml/liter	1.43	1.0	0.089
mg-atoms/liter	16.0	11.2	1.0

depths of oceans and lakes are without oxygen and life, but such is not the case. Both have been found at the greatest depths that have been sounded, but oxygen is absent in a few places and low enough in many places to limit the habitation by animals.

Oxygen occurs in the deep waters only because these waters were once near the surface. In most lakes all of the waters are near the surface during the seasonal overturn. In the sea oxygen enters the water in polar regions from which it is carried by the cold water as it sinks to great depths and flows slowly toward the equator. Because of the low temperatures and the scarce life in the depths the oxygen is used very slowly indeed. Water from the depths of the Atlantic Ocean is estimated to have been at the surface about 500 years ago. The deep water of the Pacific is about twice as old.

An outstanding feature of the distribution of oxygen in the sea is the prevalence of a layer with low oxygen content at intermediate depths—usually between 400 and 1200 m. The oxygen concentration is commonly either 5 mg/liter or higher, but over large parts of the Indian and Pacific Oceans it is below 1 mg/liter, a level that limits markedly the kinds of animals that can inhabit such waters. Although the layer with low oxygen content (called the oxygen minimum layer) is usually below the level of fishing, in regions of upwelling it may be within 100 m of the surface or even appear at the surface. A few stagnant basins, of which the Black Sea is a major example, contain deeper water that does not recirculate to the surface. Such water may contain hydrogen sulfide and no oxygen. Oxygen is usually present in abundance in all open ocean waters that can be reached by commercial fishing gear.

Severe restriction of the distribution of animals by shortage of oxygen is much more common in fresh water or in estuaries because the decomposable organic materials are much more concentrated either naturally or because of pollution. The deeper parts of such waters may be devoid of oxygen regularly during the summer. Turbulent streams will normally contain water saturated with oxygen, but in their sluggish lower reaches the oxygen may be used almost entirely for the respiration of plants each night and too little may be left for the fish. Critical shortage of oxygen also develops occasionally when exchange with the atmosphere is cut off by ice—a time when photosynthesis is minimal. The rate of respiration by plants and animals is also reduced by the cold but may remain high enough for all of the oxygen under the ice to be consumed, a situation that is called "winterkill."

The surface layer of quiet water frequently contains more oxygen than the saturation level. Excess oxygen commonly persists as the waters warm during the day but is used in respiration or lost to the atmosphere during the night. Excess oxygen in air may be harmful to land animals, but excess oxygen in the water at concentrations that occur naturally has never been found to be harmful.

2.4.2 NITROGEN

Fixed nitrogen in the form of nitrates, nitrites, or ammonia is essential for life of all kinds (see Section 2.4.5), but only a few nitrogen-fixing bacteria and plants can use the molecular form N_2. It dissolves in natural waters at atmospheric presure in amounts about 1.7–1.8 times the saturation values of oxygen. The amounts utilized by the nitrogen-fixing organisms are such a tiny fraction of the amount available that the reduction in concentration is rarely detectable in ordinary analyses. In fact, the presence of N_2 at all depths of lakes and seas in amounts near saturation levels has been used as an argument that the deep waters were once in contact with the atmosphere.

However, problems with N_2 in water arise when too much is present. Supersaturation commonly causes trouble for aquatic organisms because it leaves solution inside the body to form bubbles which remain in tissues for a long time. When human divers have this trouble it is called the "bends." Supersaturation of natural waters by N_2 occurs in situations that are of special concern to the fisheries scientist. First, emerging ground water is occasionally supersaturated and unusable without deaeration for rearing of fish. Second, water falling from natural falls or dams carries air below the surface of the plunge basin where it dissolves under the added pressure of the water above. The extra N_2 may cause trouble for fish downstream from the plunge basin. Third, water warmed either naturally or artificially may become supersaturated and harmful to fish.

2.4.3 CARBON DIOXIDE AND pH

CO_2 is present in natural waters in highly variable amounts because it reacts with the water and other dissolved materials to form either carbonic acid, H_2CO_3; bicarbonate ions, HCO_3^-; or carbonate ions, CO_3^{2-}. Apparently CO_2 is always present in adequate quantities for the biological needs of plants and never a limiting factor for animals in a natural environment because of abundance (unless pH is also low). The tolerance of animals varies greatly, but in fresh water artificial concentrations greater than about 50 mg/liter generally cause distress or death.

The measure of hydrogen ion concentration, pH, varies in natural fresh waters from a low of 1.7 in extremely acid waters to a high of about 12 in a few soda lakes. Values outside the range of about 4.5–10 are detrimental to fish but rarely occur.

The pH of the open sea, in contrast to that of fresh water, remains almost always between 8.1 and 8.3 in the surface layer. At the depth of the layer of minimum oxygen the pH may be about 7.5; in stagnant basins containing hydrogen sulfide it may be as low as 7.0. The pH is not known to be a limiting factor in the sea for animals and plants of commercial importance.

2.4.4 SALINITY

Pure water does not occur in nature. Even rain water contains dissolved gases and various salts derived from sea spray or dust. The water that is commonly regarded as fresh contains less than 1 g/kg of dissolved solids. "Hard" water contains about 300 mg/kg; exceptionally "soft" water about 40 mg/kg; average river and lake water 100–150 mg/kg. Average seawater contains about 35,000 mg/kg or 35 g/kg, a value that is commonly expressed as $35^0/_{00}$, which is the equivalent of 3.5%.

Salinity is defined as the total amount of solid material in grams contained in 1 kg of seawater when all the carbonate has been converted to oxide, the bromine and iodine replaced by chlorine, and all the organic matter completely oxidized. Seawater is a mixture of remarkable constant proportions of the halides, carbonate, and sulfate salts of sodium, magnesium, calcium, potassium and strontium, together with small quantities of other substances and minute traces of many other elements. The proportions between the major constituents are so constant that the total salinity has been estimated simply by determining the amount of chloride (plus bromide), the preponderant anion, and applying the formula:

$$\text{Salinity} \, (^0/_{00}) = 0.03 + 1.806 \times \text{chlorinity} \, (^0/_{00})$$

Recently, a formula has been developed to relate the salinity to the electrical conductivity because this property can be measured with great precision.

Such a constant proportion of salts does not occur in fresh water nor in enclosed basins with salt lakes such as the Caspian Sea, Dead Sea, or Great Salt Lake. These may contain preponderantly carbonates, sulfates, chlorides, or any possible mixture of them. The proportion of salts in estuaries where rivers mix with the ocean may be slightly modified by the salts brought in by the river.

The salinity of the open ocean surface varies only from about 33 to $37^0/_{00}$ because of the variations in the amounts of evaporation and rainfall. Salinity tends to be higher in the dry, trade-wind belts of the subtropical regions than in the regions of the moist, westerly winds. *Haloclines* (increase of salinity with depth) may exist at the bottom of the mixed surface layer (Fig. 2.10), but below the halocline the salinity varies only slightly with either depth, locality, or season. None of the variations in the open ocean are likely to be of direct biological importance.

Vastly different in magnitude and biological significance are the variations in salinity along the shores where the effects of the river flow and evaporation may be much greater. In estuaries at the mouths of rivers the salinity varies from that of fresh water to full ocean strength according to depth, location in the estuary, tide level, and the seasonal change in freshwater flow. Some coastal waters in arid climates may have salinities greater than the open ocean; $40\%_{00}$ is found in the Red Sea and Persian Gulf; $60-100\%_{00}$ in some small saltwater lagoons from which water evaporates and into which seawater enters only occasionally.

In coastal waters the differences in salinity and the rapid changes restrict all plants and animals to the salinity zones that they can accommodate (see Section 4.3). The *euryhaline* forms are those that can accommodate large changes and include most of the animals and plants that customarily live in the estuaries. Some anadromous animals, notably the salmon, migrate when young from fresh water through the estuary to the open sea and return to fresh water when ready to spawn. The *stenohaline* forms are those that tolerate little change and include those that live either in fresh water or the open sea.

2.4.5 NUTRIENT ELEMENTS

In the waters, along with the common halide, carbonate, and sulfate salts of sodium, magnesium, calcium, potassium, and strontium, occur compounds of most of the known stable elements. Most of these are in minute quantities, but many are important. In addition to all of the above, nitrogen, phosphorus, silicon, iron, manganese, zinc, copper, boron, molybdenum, cobalt, and vanadium are known to be essential for one plant or another. Others, such as nickel, titanium, zirconium, yttrium, silver, gold, cadmium, chromium, mercury, gallium, tellurium, germanium, tin, lead, arsenic, antimony, and bismuth, are concentrated by aquatic organisms, presumably for some beneficial purpose. In addition, certain vitamins and other organic compounds occur in solution and may be essential for some organisms.

Most of these elements and compounds have either functions or effects that are poorly understood. Some may be limiting factors when too much or too little is present. There is no question, however, about the role of phosphorus, nitrogen, and silicon, which are commonly called the *nutrient elements*. These must be present as soluble salts for plants to grow. Phosphorus and nitrogen go through cycles of incorporation into organic compounds by the plants and animals and subsequent breakdown into soluble salt after discard or death. Phosphorus is the most important to the ecologist because it is the most likely to be deficient and is usually concentrated, that is, needed by plants in proportions greater than any other element.

Nitrogen is available in abundance as a gas. The gas is readily soluble in water, but only certain bacteria and plants, sometimes in association with each other, can "fix" nitrogen into a form of ammonia (NH_3), nitrite (NO_2), or nitrate (NO_3) that can be used by plants. In one of these forms the nitrogen can be incorporated into the bodies of plants and used in amino acids or proteins. The plants may be eaten by herbivorous animals and in turn these are eaten by carnivorous animals. In each step of this conversion the nitrogenous compounds are first digested and then either egested or assimilated. If assimilated they are either made part of the body or excreted. The parts that are egested or excreted and the dead bodies are decomposed by bacteria back to ammonia, nitrite, and nitrate (see Chapter 5).

Phosphorus is available only from certain rocks and deposits on earth from which it enters the rivers and ground water as a phosphate ion. In this form it is usable by plants principally in the formation of proteins and fats. Like nitrogen phosphorus goes through a cycle of plant to animal to bacteria as it is built into organic compounds and broken down again to the inorganic form. Unlike nitrogen a large part of the phosphorus is quickly absorbed at the surface of the mud, in shallow waters of seas and lakes. It is restored to organisms by various processes which are not well known. Such deposition on the mud is not part of the cycle in parts of the seas where depth is greater than about 250 m.

Silicon is abundant on earth but not readily soluble. It is not required for nutrition in the strict sense of the word but is needed by certain phytoplankton for their skeletons. Apparently it is recycled rapidly from organic to soluble inorganic form, but the chemistry is not well understood. A shortage has been shown to limit phytoplankton populations, but unlike nitrogen and phosphorus it is usually available in adequate quantities.

All three of these nutrient elements usually occur in the surface waters of lakes and the ocean in small quantities which are rapidly changeable depending on their utilization by plants. Phosphorus is usually the least abundant; only 10–30 mg/m^3 of inorganic phosphate are available at the surface of many uncontaminated lakes; only 1–5 mg/m^3 in large parts of the ocean surface. Nitrogen as nitrate is more variable in amount; between 0 and 500 mg/m^3 is present in surface waters. Such small quantities and wide variability pose special problems. The quantities may be too small to be detected by ordinary laboratory chemical techniques. The variability from place to place and time to time may require large sampling programs and interpretation of the results may be difficult.

Much more significant from the standpoint of productivity than the amount of nutrients measurable is the rate at which they are supplied to the surface layers. Phosphorus as phosphate at depths of 100–1000 m in the oceans is present in amounts about ten times greater than the surface concentration of 1–5 mg/m^3. Even greater quantities of nitrates are usually found in the same intermediate water layers. High concentrations of nutrients also develop in the deeper waters of lakes or in shallow seas where some nutrients may be absorbed by the mud or deposited in a layer on the mud.

These benthic nutrients were at one time part of the biological cycles in the surface layers. There is a net downward flux of organic matter which decomposes as it sinks and the elements are returned to inorganic form at some depth in the lake or ocean. They are brought back into the biological cycle in the surface layer by physical processes such as upwelling, turbulence, and convection currents. The places where nutrients are brought up are enormously more productive than other parts of the ocean (Fig. 2.16). In lakes the restoration occurs during the seasonal overturn of the waters. (See Section 2.6.)

High concentrations of the nutrient elements are not always desirable. Phosphates and nitrates are abundant in domestic sewage wastes and in the drainage from farmlands. They frequently cause unwanted blooms of plankton in rivers, lakes, and estuaries, so we try to control the amount in our waste waters. It is an ever increasing problem to put them where they will produce desirable results.

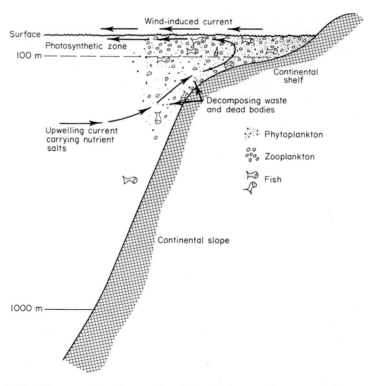

Fig. 2.16.   Diagram of the bio–geo–chemical cycle in upwelling areas of the ocean.

2.5 Photosynthesis

Carbon dioxide, water, chlorophyll, nutrient elements, trace elements, and the energy of light are converted to living protoplasm in plant cells with the accessory production of oxygen. In a grossly simplified form:

$$6\ CO_2\ +\ 6\ H_2O\ +\ \text{energy from sunlight}\ =\ C_6H_{12}O_6\ +\ 6\ O_2$$

In the absence of light all cells, either animal or plant, respire with the consumption of oxygen and the release of energy and carbon dioxide:

$$C_6H_{12}O_6\ +\ 6\ O_2\ =\ 6\ H_2O\ +\ 6\ CO_2\ +\ \text{energy}$$

Because the waters of the earth cover about 71% of the surface, they are potentially the site for about the same percentage of the earth's photosynthesis. Whether the waters do support most of the earth's photosynthesis is not known because the plants are vastly different from those on land, although we guess that photosynthesis is at about the same rate per unit area. Practically all of the plants in the waters are the minute, pelagic phytoplankton that drift throughout the surface layer. The attached algae and higher plants that occur only in shallow water along the shores comprise but a minute part of the total weight of aquatic plants, although they may be important in marshes, small lakes, or streams.

The photosynthetic zone in the waters may be determined by observing the distribution of attached plants on the bottom since they can live only where they receive adequate light. Such direct observations are not satisfactory for use in the open water because some of the drifting phytoplankton sink below the photosynthetic zone and may survive there for a time.

Instead, samples of the water containing the phytoplankton are taken from various depths; each is divided in two parts, one of which is placed in an opaque bottle, the other in a clear glass bottle. Both are incubated at the temperature of the water from which the samples were taken, and the clear bottle is exposed to light of the intensity at the depth at which it was taken. After incubation the amounts of photosynthesis or respiration are estimated by measurement either of the changes in the oxygen content of the bottles or of the changes in the carbon content of the plants. The latter change is measured by adding a small known amount of radioactive carbon-14 to the bottle before incubation and measuring the radioactivity of the plants after incubation (see Chapter 5).

Such methods have shown that the amount of photosynthesis that occurs in the water is enormously variable. There are many kinds of phytoplankton which occur in different concentrations, and they react differently to light of different colors, to different temperatures, and to different light intensities. In general, however, the maximum amount of photosynthesis occurs in the water layer receiving between 70% and 30% of the light at the surface regardless of the temperature, depth, actual amount of light, and whether the water is fresh or salt. Less photosynthesis occurs in the brightly lighted surface; less in the dimly illuminated depths below (Fig. 2.17).

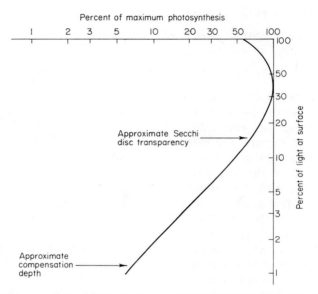

Fig. 2.17. Diagram of distribution of photosynthesis in the surface layer. [Adapted from W. Rodhe (1965). Standard correlations between pelagic photosynthesis and light. *Mem. Inst. Ital. Idrobiol.*, **18**, Suppl., 365–381.]

An important indicator is the *compensation depth*, the depth at which photosynthesis is just balanced by respiration. It is usually found at approximately the depth reached by 1% of the incident light although it varies with temperature and the kind of phytoplankton. A convenient approximation is that the compensation depth is about three times as great as the Secchi disc transparency.

The total amount of photosynthesis under a unit area of surface varies according to light, nutrients, temperature, and kinds of plankton. In the sea the amount of photosynthesis varies markedly from place to place and serves as a useful index of the ability of various regions to produce fish. The range is about tenfold, and the regional minima occur in the clear subtropical oceans and the regional maxima in two kinds of locations of special importance to the fisheries (Fig. 2.18), one on the west coasts of the continents and the other in the subarctic areas.

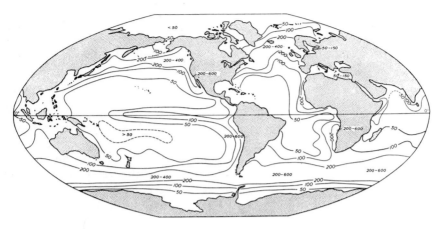

Fig. 2.18. Estimated annual photosynthetic productivity of the oceans. Units: grams of carbon/meter²/year. [Source: R. H. Fleming. *In* Hedgpeth (1957), pp. 87–107.]

2.6 Circulation

Currents are always present in the waters, even in seemingly quiet lakes. The currents transport heat, dissolved materials, and solids. The heat influences the climate over the waters and the surrounding land, the dissolved materials sustain the life of the waters, and the solids determine the kind of bottom. Ocean currents flowing from the subtropical latitudes warm great areas of the northern Atlantic and the northern Pacific Oceans. Vertical currents bring not only the cool waters to the surface but also nutrient materials from the depths of lakes and the sea. All of the currents pick up, carry, sort, and deposit nonliving particles in the water to form the bottoms and the shores.

Organisms adapt either body or habit in many ways to live with the currents. Bodies are shaped for swimming rapidly or slowly. Many animals and plants are equipped with attachments to suit the bottom and velocity of turbulent waters in streams or on wave-pounded beaches. Spores, eggs, and larvae may drift with the currents in the open sea for hundreds of kilometers. Some animals depend on currents to bring food to them. Some whales and fish may ride ocean currents for 10,000 km in their regular migrations.

Of great interest to the fisheries are the currents that concentrate and sustain the animals. The tidal currents bring food to oysters and cause fish to concentrate where they are easy to catch. The great upwellings along the west coasts of the continents and the large plant production sustain the vast shoals of the herringlike fishes. The poleward extensions of the warm currents make habitable for many fishes large areas of the oceans. They even include the current in the stream that sweeps the angler's fly into the fish's mouth.

2.6.1 CAUSES OF CURRENTS

Water currents are the results of a complex of forces most of which receive their energy from the sun either directly or indirectly. Many of the forces are almost vanishingly small, but their effects are substantial as the balance among them changes. Wind drag on the surface of the water can pile it up slightly along the shore. Atmospheric pressure commonly varies by the equivalent of between 10 and 20 cm of water and occasionally by more than 50 cm. Water changes its volume at the surface as it is either heated or cooled and as it either receives rain or evaporates. The gravitational attraction of the sun and moon pulls the water toward them in regular daily and lunar cycles to cause the tides. All of the resulting paths of water movement are affected by the rotation of the earth. Finally, all water masses in motion have inertia and very little friction, so they tend to have long-lasting oscillatory motions.

The *wind* is the primary driving force of the surface currents in both lakes and oceans. As the wind increases in speed, it accelerates the water by the shearing stress it exerts on the surface; and the surface water in turn exerts a lesser shearing stress on the water layers below. As the wind drags the water across a lake it lowers the water level slightly on the side toward the wind and raises it on the far shore. From the far shore the water will tend to flow back as a gradient current either along a part of the lake less exposed to the wind or in the lower part of the mixed surface layer. In the open sea the wind may push the water for thousands of miles at varying speeds, and the gradient current may be noticeable along only the coast on which the wind blows.

The effects of wind stress on water prevail only in the upper layers—principally in the mixed layer at the surface. In the thermocline layer and in the deeper waters the wind-driven currents can have only a secondary effect through piling up or moving away the upper layer. Here the currents are sustained largely by differences in density.

Coriolis force, or the effect of the rotation of the earth, affects any water in motion. It causes the water to be deflected toward the right in the northern hemisphere and toward the left in the southern hemisphere. This deflection in wind-driven currents was predicted from theory by Ekman to be 45° at the surface over deep water in high latitudes and to increase in angle with depth (Fig. 2.19). The theory appears to be approximately correct, but it has

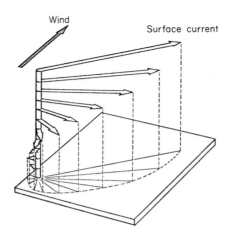

Fig. 2.19. Diagram of the direction and velocity of water driven by the wind in the northern hemisphere. The change of direction with depth is called the "Ekman spiral." [After Sverdrup, Johnson, and Fleming (1942).]

never been verified exactly. The deflection is less, however, in water shallower than about 100 m, less in low latitudes, and less at low wind velocities.

Changes in *density* arise at the surface as the water is either warmed or cooled and as the sea either receives rain or evaporates. The changes extend throughout the mixed layer above the thermocline because of wind-caused turbulence. The surface layer will sink when it becomes more dense than the deeper water during the seasonal temperature changes in lakes and the poleward movement of water in the ocean. Once below the surface a water mass remains almost constant in temperature and salinity; these properties are called "conservative" properties since they are changed only slowly by mixing processes. The masses of water that sink in the colder oceans move slowly toward the equator, so slowly that some are estimated to have been at the surface 2000 years ago!

However, neither wind nor density is the cause of currents so obvious to everyone living near where *tides* occur. Tides along ocean coasts and in estuaries have vertical ranges generally between 1 and 5 m and, in certain places, more than 8 m. These periodic changes in sea level are associated with periodic currents in estuaries and over the continental shelves. Tidal currents are usually much stronger than other movements. Their speeds and direction depend not only on the rise and fall of the tide but also upon the depth and configuration of the bottom. Maximum tidal currents may exceed 5 m/second (10 knots) in some narrow straits, and tidal currents of 0.5–1 m/second are found commonly on continental shelves and in estuaries.

Such currents on the continental shelves are rotating currents that result in nearly zero transport of water throughout a tidal cycle. They do, however, have a major influence on the bottom composition of the shelves and sometimes enhance the upward transport of nutrient materials through the turbulence that they cause.

Except along the exposed sea coasts and their estuaries, tides and tidal currents have little effect (so far as is known) on either circulation or fertility of the waters. Tides in the open sea seldom range more than 1 m in height. Tides in the larger lakes are separable either from oscillatory motion or from tides in the earth's crust only with considerable uncertainty.

In enclosed or nearly enclosed basins, such as estuaries, inland seas, or lakes, the water surface tends to oscillate after having been moved by any force like water sloshing up and down in a tub. Such oscillations are *seiches*. Their amplitude depends on the original force; their frequency on the size and shape of the basin and on the number of harmonics. In simply shaped basins the frequency is predictable from formulas. Lakes of about 10 km in length have been found to have principal oscillatory periods ranging from 12 to 32 minutes and shorter period harmonics.

Not only may a lake surface oscillate, but also the layers of different densities may oscillate relative to one another. Such *internal seiches* may be evidenced by a periodic rising and falling of the thermocline. These internal seiches, although not well understood, have a much greater period and amplitude than surface seiches; therefore they probably have a greater influence on bottom sedimentation and in transport of nutrients from the bottom to surface layers.

2.6.2 WAVES

As everyone knows the wind causes waves on the surface of the water, small ones at first and then larger ones as it increases in speed. These waves progress with the wind at speeds much faster than the water currents set up by the wind which are commonly only about 1 or 2% of the wind speed. The movement of a water particle in a nonbreaking wave is nearly orbital with a maximum speed in the direction of the wave movement in the wave crest, a maximum speed in the opposite direction in the wave trough, and nearly vertical motion when the water is at its mean level (Fig. 2.20).

The speed of a wave in deep water (not the net current) is related to its wavelength; its speed in centimeters per second is approximately 12.5 times the square root of its wavelength in centimeters. In water shallower than one-half of the wavelength the wave slows down to become a function of depth and when it slows sufficiently it begins to break. Since the speed of the wave varies according to the wavelength, the long storm waves travel faster than the shorter waves. Long wavelength waves travel thousands of kilo-

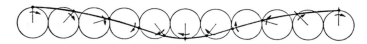

Wave travel

Fig. 2.20. Diagram of the direction of water movement in a progressive wave.

meters with little attenuation, so the common state of the ocean's surface is a mixture of *swell*, the long wavelength waves generated by a distant storm, and *sea*, the shorter waves produced by the local wind.

Most density gradients in natural waters also have waves. Such internal waves are usually found in the thermocline. They tend to be larger than surface waves but travel much more slowly. Some associated with diurnal tides have a period of about 12 hours and an amplitude of between 30 and 50 m. Others have periods of between 15 and 30 minutes and amplitudes of about 6 m. Their cause is not well known. Sometimes they can be seen on the surface when they cause parallel slicks or bands of alternating rippled and smooth water. They are usually measured by repeated determination of the depth of the density gradient (thermocline) at a fixed point (Fig. 2.21).

Such waves probably enhance the vertical circulation of nutrients in some places, but they are of interest to fishery scientists primarily because they are associated with seiches which are the cause of the principal subsurface currents in lakes. Waves also cause substantial variation in the depth of the thermocline and interfere seriously at times with the interpretation of echos from echo-ranging equipment.

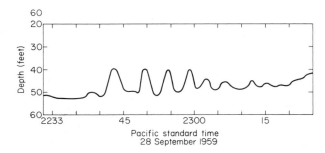

Fig. 2.21. Small internal waves off San Diego were evidenced by the changing level of the 67°F isotherm. These waves moved shoreward at a speed of about 10 cm/second. [After O. S. Lee (1961). Observations on internal waves in shallow water. *Limnol. Oceanogr.* **6** (3), 312–321.]

2.6.3 MEASUREMENT OF CURRENTS

A major problem for all who are concerned with the circulation of the waters is measurement of the speed, direction, and volume transport of the currents. Speed past a fixed point is measured directly where possible by current meters, some of which may record time and direction of flow. Trajectories may be traced by observing the drift of floating bottles, poles, cards, dye, or buoys. The latter may be equipped with radio in order to be tracked by direction-finding radio receivers. Also, a ship drifts with the surface current, which may be estimated by the deviation of the ship from its projected course.

Such direct methods are used effectively in rivers, lakes, and estuaries where instruments can be easily anchored or where numbers of drifting objects can be easily observed. They are, however, of limited usefulness in the open sea or in large deep lakes where anchoring is difficult and where it is very expensive to follow drifting objects. Here it is possible to use an indirect method to estimate the complete field of motion in a body of water. This is the dynamic or geostrophic method based on measurements of the internal distribution of pressure.

The pressure field cannot be measured directly by pressure meters with adequate accuracy but can be estimated from the density of columns of water ending at the surface. The density is determined from very accurate measurements of temperature and salinity at various depths along the length of the column. The necessary temperature and salinity determinations are obtained by "hydrographic casts" in which a series of collecting bottles with reversing thermometers are lowered on a wire to a depth usually of 1000 m or more. These are closed and reversed when at depth and retrieved with the temperature reading and a water sample from which salinity and other determinations may be made. Such casts are made in a series or network of stations.

This method of estimating currents is based on the assumption that the currents are steady, a state that is approximated in fact to a remarkable degree by the great ocean currents. These are induced and modified by winds, atmospheric pressure, change in density, coriolis force, and gravity, all of which are relatively tiny forces acting on a very large mass with great inertia. The major currents are little affected by tides or other waves.

A full explanation of the dynamic method requires mathematics that cannot be introduced here. The explanation that follows is a grossly simplified approximation.

The method is based on a measurement of the difference in pressure gradient from which can be computed the difference in current velocity between two surfaces or horizons in the sea. If one of these surfaces has no motion or if its motion is known then the actual motion of the other surface can be computed. Usually the surface of no motion is assumed to be at a depth of 1000 m, an assumption that has been found to be reasonable in many localities.

The determination of density from temperature and salinity for a water column between a lower surface and a higher surface (usually the air–water surface) permits a determination of the thickness of the intervening layer. This is expressed in terms of the work done per unit mass against gravity, but the unit has been chosen as $D = 10^5$ dyne-cm/g which is very nearly equivalent to a linear meter. It is called a dynamic meter. Since only the difference in thickness is needed for the computation of currents, the results are expressed conveniently as dynamic height differences from a standard water column at 35‰ salinity and 0°C, water slightly more dense than that ever found at sea. The differences found in practice are commonly a few dynamic centimeters.

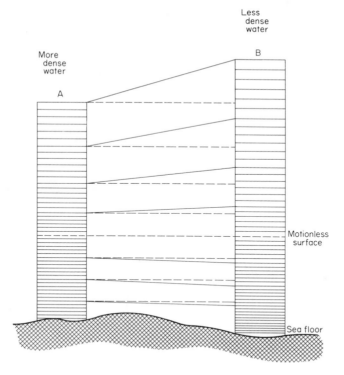

Fig. 2.22. Diagram of two water columns with a different density distribution above and below a layer of no motion. The solid lines indicate equal pressure, the dashed lines a level surface. The *less* dense water in the upper part of column B is under *more* pressure than the water at the same level in column A.

As an illustration of the principle of measuring dynamic heights consider two points A and B, at which the density of a deep column of water is determined (Fig. 2.22). The less dense water above a surface of no motion is at B; hence the air–water surface at B is higher. The pressure is equal in A and B at the surface of no motion indicated by 0, but above this surface in the B column the pressure is higher than at the same level in A. Since the basic condition is a steady current that maintains the difference in elevation, the water does not flow directly downhill but rather flows in the direction induced by the coriolis force at right angles to the slope.

When the dynamic height is determined at a network of stations it is possible to compute a dynamic topography (Fig. 2.23A) and to estimate the current direction and velocity (Fig. 2.23B). The current direction is clockwise around an elevation and counterclockwise around a depression in the northern hemisphere. Thus if an elevation is on an observer's right the

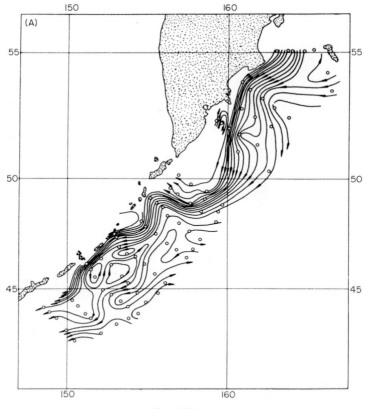

Fig. 2.23A

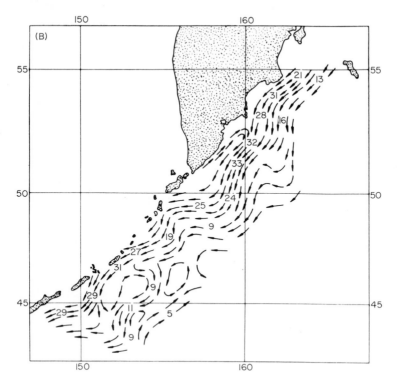

Fig. 2.23. (A) Dynamic topography of the sea surface estimated from determinations made at the stations indicated by circles. Contours are spaced 2 dynamic centimeters [Source: L. M. Fomin (1964). "The Dynamic Method in Oceanography." Elsevier, Amsterdam.] (B) Currents computed in centimeters/second from slope of dynamic surface. [Source: L. M. Fomin (1964). "The Dynamic Method in Oceanography." Elsevier, Amsterdam.]

direction of flow is away from the observer. The velocity is higher along steeper slopes and can be predicted with reasonable accuracy from the gradient of the slope.

Another indirect method has been developed around the measurement of the electric potential produced by the highly conductive seawater moving through the magnetic field of the earth. The gradients in electrical potential (see Section 2.3.5) are measured with a geomagneticelectrokinetograph (GEK). When suitable measurements are made along different compass courses, the direction and speed of the surface currents may be estimated with fair accuracy.

No method is satisfactory for all purposes. The indirect methods work best in the open sea where the assumption of a steady state may be approximately correct and where short period changes are less important. The direct methods work best in estuaries, lakes, and on continental shelves where tidal and other oscillatory currents preponderate. Measurement of deep currents and vertical currents is especially difficult and expensive.

The direction of water flow is always reported as the downstream motion; e.g., an easterly current moves water toward the east. This practice is opposite to that of reporting wind direction; e.g., an easterly wind blows from the east.

Volume of flow can be estimated after determination of the velocity over a representative cross section of the current. The simplest situation is in a stream in which volume is estimated from width, average depth, and average velocity. Measurement of the volume of ocean or lake currents is done in fundamentally the same way, but determination of the boundaries and average velocity is much more complicated.

2.6.4 CIRCULATION IN LAKES

The surface waters of lakes are moved to and fro by the wind. The resulting currents are always turbulent and depend on many factors including the shape and depth of the lake, and strength, steadiness, and fetch (length of water stressed by the wind) of the wind. Consequently they are highly irregular and largely unpredictable. What needs emphasis here is the disturbance of the deeper waters which reflects in part the disturbance of the surface and which may be significant to the fisheries.

When the wind piles up the water at the downwind end of a large lake, the thermocline is also tilted downward at the downwind end. A pile up of a few centimeters at the surface may produce a depression of the thermocline of several meters and the consequent displacement of a large volume of deep water below the thermocline. The pile up includes the warm surface water, the removal of which exposes cooler water at the upwind side, thus enhancing the mixing of the surface layer. The return gradient currents of the surface layer may be located just above the thermocline and will be accompanied by parallel currents in the layer just below the thermocline (Fig. 2.24). Since

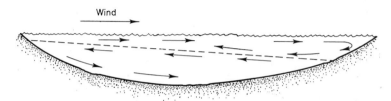

Fig. 2.24. Diagram of a vertical section of a lake, showing circulation above and below the thermocline (- - - - - - -).

the lake basin is never symmetrical, the currents all tend to rotate. They are never either laminar or smooth flowing but turbulent and accompanied by waves, both surface and internal. Sometimes the currents reach surprisingly high velocities; surface current averages about 2% of wind speed. When the wind dies the currents and waves do not stop but continue to oscillate in various ways. The movement is analagous to the sloshing of water in a tub, but the changes in height of the surface layer are slight, the changes in height of the thermocline substantial. Such currents are of special importance to fisheries when they displace upward or downward a cold layer, a low oxygen layer, or a nutrient-rich layer.

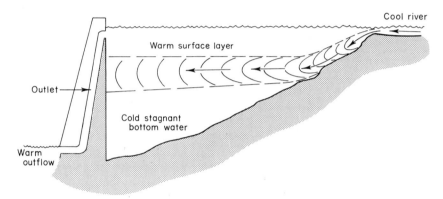

Fig. 2.25.   Diagram of distribution of tributary water in a reservoir during the summer.

Another significant circulatory pattern develops in lakes with large tributaries having a different density from the lake surface because of either temperature or silt. If the inflow is lighter it will spread over the surface; if heavier, it will descend and intrude in a layer at the depth where it finds its density (Fig. 2.25). The latter condition is common in reservoirs with cold tributaries. If the river water includes organic material that uses up the oxygen, the layer of water may be a barrier to the movement of fish.

2.6.5 CIRCULATION IN ESTUARIES

Estuaries are mixing basins of fresh and salt water. The circulation is much more rapid than in either the lakes or open ocean. No two estuaries are alike and the mixing is as varied as the tidal range, the river flow, the shape, size and depth of the estuary, the wind, and the effect of the coriolis force. Thus the circulation of each estuary under study needs to be individually determined. If detailed information under varied conditions is needed, it may be necessary to build a hydraulic scale model.

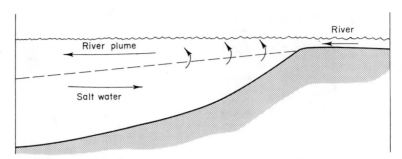

Fig. 2.26. Diagram of water flow in an estuary above and below the halocline (--------). The volume of inflowing salt water is usually several times as great as the volume of inflowing fresh water.

The mixing in each estuary occurs as the fresh water flows out on the surface and the salt water flows *in* below and mixes with the fresh water above because of turbulence (Fig. 2.26). The relative inflow of salt water with respect to fresh water varies according to the depth and the amount of mixing, but in large, deep estuaries the saltwater inflow may be between 10 and 20 times as great as the freshwater inflow. The mixture of fresh and salt water flows out on the surface. The actual flushing pattern can be determined for an estuary, and often effluents can be introduced where they will flow rapidly out to sea and not circulate in the estuary.

Another aspect of the mixing is the horizontal and vertical incursion of the fresh- and saltwater layers. The incoming tides may push relatively unmixed seawater far up along the bottom of a river; the outgoing tides may take relatively unmixed fresh water far out on the surface of the estuary. These limits will be also the limits for animals that can tolerate only salt water and for those that can tolerate only fresh water.

The tides affect not only the area where fresh and salt water mix but also the fresh water upstream. Normally the fresh water rises and falls with the tide upriver to the zone where the riverbed is approximately at the elevation of the high tides. Toward the sea from this zone will be a zone in which the fresh water will just stop flowing at high tide, and still further seaward will be a zone in which the fresh water will flow back and forth with the tides.

2.6.6 SURFACE CIRCULATION IN THE OCEANS

Knowledge of ocean currents has been important to navigators who have sailed the oceans. They have learned about the dangerous currents and the currents that can help them along. Most of this knowledge has concerned the coastal currents, many of which are caused by the tides. Extensive charts and tables showing the strength and direction of such currents are available.

Not nearly as well known in detail are the nontidal currents of the open ocean which are driven by wind and differences in density of the water. These include the great ocean rivers such as the Gulf Stream which have been known to navigators for centuries but little understood and poorly charted. These are the currents that are of importance in the transfer of heat around the globe and to the life in all oceans. They are the subject of this discussion.

The circulation of the world oceans is dominated by the five great gyres: North Pacific, South Pacific, North Atlantic, South Atlantic, and southern Indian Oceans (Fig. 2.27). These are all clockwise in the northern hemisphere, counterclockwise in the southern hemisphere. All have easterly flowing currents driven by the west winds at about 40° or 50° latitude North or South, and westerly flowing currents near the equator driven by the northeast or southeast trade winds. All have currents flowing toward the pole in western oceans, towards the equator in the eastern oceans.

Off these great gyres spin smaller gyres in neighboring gulfs and seas. The Gulf of Alaska, Tasman Sea, and the Greenland Sea all have connected gyres. More isolated waters such as the Bering Sea, South China Sea, Chukchi Sea, and Gulf of Mexico seem to have independent gyres. Almost

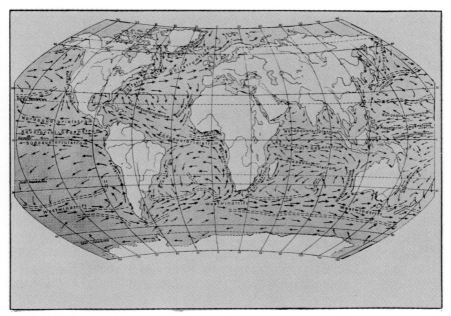

Fig. 2.27. Major currents of the oceans. [Source: U. Scharnow (1961). "Ozeanographie für Nautiker." Transpress, Berlin.]

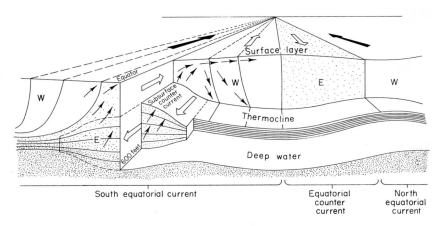

Fig. 2.28. Diagram of a cross section of the Pacific Ocean about 1500 km long at 150°
W long showing the major horizontal currents. [Source: O. E. Sette *et al.* (1954). Progress
in Pacific oceanic fishery investigations, 1950–53. *U.S. Fish Wildl. Serv., Spec. Sci. Rep.–
Fish.* (116), 75.]

every basin in every ocean contains a gyre which may or may not turn in
the same direction as the major gyres.

Between the great gyres of the northern and southern oceans are the
countercurrents which flow toward the east near the equator. One of these,
a surface current in the Pacific at about 8° North latitude, has been well
known for centuries. But recently a weaker surface countercurrent south of
the equator was discovered and more important a powerful submerged
current, the Equatorial Undercurrent or Cromwell Current, was also discovered. Both surface countercurrents and undercurrents occur also in the
Atlantic and Indian Oceans.

The discovery of the Pacific Equatorial Undercurrent by Townsend
Cromwell of the U.S. Bureau of Commercial Fisheries illustrates how little
is known of subsurface ocean currents and indicates something of the
opportunities remaining in oceanography. Totally unpredicted by oceanographic theory the presence of the current was suspected from the easterly
drift of fishing gear, and its velocity was measured by Cromwell in 1952.
Further studies by Cromwell and others revealed a persistent, ribbonlike
current about 300 km wide lying only between about 70 and 200 m below the
surface, at least 6500 km long, with a speed at the core of about 1.5 m/second
(Fig. 2.28). It transports about 40×10^6 m³/second,* a volume half as great

---

*1 × 10^6 m³/second is about 35.3×10^6 feet³/second. The unit is called a "Sverdrup"
(sv) by some oceanographers.

as the Gulf Stream or nearly 200 times as large as the Amazon River at flood stage.

The greatest current in the world is the west wind drift; it flows around Antarctica in an unbroken circle to form part of each of the three great gyres of the southern oceans. The transport of water has been estimated to be $150-190 \times 10^6$ m^3/second.

Two of the best-known ocean currents are the Gulf Stream and the Japan Current or Kuroshio. These currents flow northeasterly in the northwest Atlantic and Pacific, respectively. Each of these currents is relatively narrow in the western oceans and then becomes broader as it crosses the ocean. Each brings warmer water from the subtropical ocean to the northeastern oceans to influence the climate of Europe and northwestern North America. Each also provides fish abundantly—tunas and other subtropical species in the warmer, western parts, herring and bottom fishes in the cooler, eastern parts. Each is connected with the north equatorial currents by diffuse currents flowing south in the eastern oceans.

Two other well-known currents play major roles in the climate and fisheries of the southern hemisphere, but they flow northward in the eastern parts of the Atlantic and the Pacific. The Benguela and Peru or Humboldt currents are off southwest Africa and western South America, respectively, and bring cool water northward toward the equator. Both are important in cooling and drying the climate of the coasts. Both support huge fisheries primarily for fish of the herring and anchovy families.

Many other currents are persistent and defined well enough to be named. Each forms but a part of a circulation system and each is significant to the life of the ocean of which it is a part.

No current is a smooth, straight-sided flow, but all tend to be turbulent and all tend to twist, that is, they include some vertical flow as well as horizontal flow. Currents converge, and when they do the less dense water flows over the more dense and the homogeneous surface layer becomes thicker. Elsewhere currents may diverge, in which case the surface layer becomes thinner and nutrient-rich water rises closer to the surface. Both convergences and divergences may be either semipermanent features of major circulation or temporary phenomena of giant eddies.

Convergences bring together plankton animals of the surface layer. When these plankton animals resist sinking, they accumulate along the convergence zone and provide food for fish. Sometimes a convergence brings together flotsam in a line on the surface, such as is commonly seen in estuaries. Whether marked by debris or not convergences are boundaries of water masses that may differ substantially in temperature and salinity.

Divergences are especially important for their role in bringing nutrient-rich water closer to the surface. They occur especially off the west coasts of Europe, Africa, North and South America; that is, they occur on the eastern sides of both northern and southern oceans. As the winds push the water toward the equator, the coriolis force sets the currents to the right in the northern hemisphere and to the left in the southern hemisphere—offshore in each case. The consequence is transport of cooler nutrient-rich water toward the surface.

Both convergences and divergences occur in the equatorial circulation (Fig. 2.29). Exactly at the equator and at the boundary of the north equatorial current and the equatorial countercurrent the westerly flowing currents tend to diverge. At the boundary of the south equatorial current and the equatorial countercurrent they converge.

In the southern oceans two parallel convergences occur in the west wind drift (Fig. 2.30). The most southerly, the Antarctic Convergence, is the zone at which the colder, fresher Antarctic water from the melting ice tends to flow beneath water about 4°C warmer. Then about 10° latitude further north is the Subtropical Convergence, the boundary of the intermediate water and the much warmer water to the north. Similar but less well-defined convergences appear near the west wind drift of the northern oceans.

Charting the ocean currents and their variations in position and strength remains one of the challenging tasks of oceanography. Major advances have been made during the 1960's in better understanding of the Indian and Arctic Oceans. Major difficulties persist, however, in the great cost of ocean observations, in the need for deep-water observations which require highly sophisticated equipment, and in the great complexity of the current systems.

2.7 Summary—The Environment of Fish and Fishing

Let us imagine how it would be to live in a permanent, thick, cold fog through which no eye can see more than 50 m but through which sunlight can penetrate for several hundred meters. Imagine next going to the top of a hill emerging from the fog and being blinded by the bright sunshine. Imagine being able to smell all of the odors from the other organisms and waters that pass you in the veil. Imagine hearing every animal near you and even the echos of your own noises reflected back to you. Imagine also the sensation of resting with little effort; always with the sensation of weightlessness but always fighting the water to move. Imagine traveling horizontally in the open sea for hundreds of kilometers with no perceptible change in climate but finding intense cold and darkness when descending only a few meters. Imagine the constant risk of sinking down into the dark and cold or rising up to the surface if depth controls ceased to function. Such a

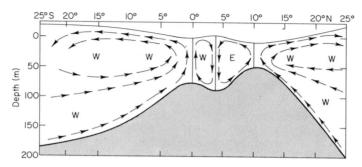

Fig. 2.29. Schematic representation of the vertical circulation within the equatorial region of the Atlantic. The direction of the currents is indicated by the letters W and E. The water below the discontinuity surface, which is supposed to be at rest, is shaded. [Source: Sverdrup, Johnson, and Fleming (1942).]

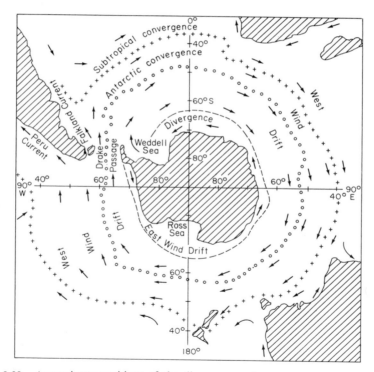

Fig. 2.30. Approximate positions of the divergence and convergences of the southern oceans. [Source: G. L. Pickard (1963).]

fancy gives a remote concept of an animal in the water in terms of human senses.

Now let us look at the environment as a whole. Perhaps the outstanding characteristic is the remarkable similarity over long distances of very thin layers that are greatly different from other layers above and below. The pressure, light, heat, oxygen, and nutrient elements all vary greatly with depth, and each is vital to the life in the water. Other features, such as N_2, CO_2, pH, density, and salinity, vary so little with depth that living things are not affected directly although the slight variations are important for physical reasons.

Another feature is the nearly perpetual motion—mostly horizontal but occasionally and importantly vertical. The currents are always carrying heat, food, eggs, larvae, and the plants and animals themselves. The organism must breast the current or change depth to a favorable current if it is to avoid transportation.

As judged by the scant number of species the more formidable of aquatic environments are not only the extremely hot, the extremely cold, and the low oxygen parts of the oceans but also the most variable parts, the estuaries. Here the currents, salinity, and temperature change in tidal and seasonal cycles in ways that few organisms can withstand, and here too most of man's wastes discharge to start their final dilution.

REFERENCES

In most instances, people studying the sea have worked independently of those studying either the fresh water or the estuaries. Consequently, the literature is comprised of either marine or freshwater treatises. We are fortunate to have a number of excellent compendia:

Chorley, R. J., ed. (1969). "Water, Earth, and Man." Methuen, London.

Hedgpeth, J. W., ed. (1957). "Treatise on Marine Ecology and Paleoecology," Vol. 1. Ecology. *Geol. Soc. Amer., Mem.* **67**.

Hill, M. N., ed. (1962). "The Sea. Ideas and Observations on Progress in the Study of the Seas," Vol. 1. Physical Oceanography. Wiley, New York.

Hill, M. N., ed. (1963). "The Sea. Ideas and Observations on Progress in the Study of the Seas," Vol. 2. The Composition of Sea-Water. Comparative and Descriptive Oceanography. Wiley, New York.

Hutchinson, G. E. (1957). "A Treatise on Limnology," Vol. 1. Geography, Physics, and Chemistry. Wiley, New York.

Lauff, G. H., ed. (1967). "Estuaries," Publ. No. 83. Amer. Ass. Advance. Sci., Washington, D.C.

Rigby, M., ed. (1965). "Collected Bibliographies on Physical Oceanography (1953–1964)." Amer. Meterol. Soc., Washington. D.C.

Sverdrup, H. U., Johnson, M. W., and Fleming, R. H. (1942). "The Oceans. Their Physics, Chemistry, and General Biology." Prentice-Hall, Englewood Cliffs, New Jersey.

Easily readable descriptive summaries of oceanography are:

Coker, R. E. (1962). "This Great and Wide Sea." Harper, New York (originally published in 1947 by the University of North Carolina Press).

Laevastu, T. and Hela, I. (1970). "Fisheries Oceanography." Fishing News (Books) Ltd., London.

Pickard, G. L. (1963). "Descriptive Physical Oceanography." Pergamon Press, Oxford.

Shepard, F. P. (1959). "The Earth Beneath the Sea." Johns Hopkins Press, Baltimore, Maryland.

Data processing and analytical methods are given in:

Barnes, H. (1959). "Apparatus and Methods of Oceanography." Part 1. Chemical. Allen & Unwin, London.

LaFond, E. C. (1951). "Processing Oceanographic Data," Publ. No. 614. U.S. Navy Hydrogr. Off., Washington, D.C.

Strickland, J. D. H. (1960). Measuring the production of marine phytoplankton. *Bull., Fish. Res. Bd. Can.*, **122**, 1–172.

Strickland, J. D. H., and Parsons, T. R. (1960). A manual of sea water analysis. *Bull., Fish. Res. Bd. Can.*, **125**, 1–185.

Food Chain and Resource Organisms

The fishery scientist who works to enhance the production of the living resources of the waters must know the animals and plants with which he deals. He must identify them and know the relationships among them. He must recognize the major groups of which there are relatively few and know as much as possible about the minor groups of which there are many thousands. The tasks of identification and classification may be either an essential preliminary to other studies or a major task for a lifetime.

3.1 Identification and Nomenclature

Every scientific study of an animal or plant should include a correct identification with the scientific name which is based on previously published descriptions of either the organism or related organisms. The scientific name is composed of two parts, the generic and specific names, or three parts if subspecies have been described. The words in the names are latinized regardless of the language or alphabet of the study and are frequently descriptive of a significant feature of the organism. The names and their usage are governed by the International Code of Zoological Nomenclature and by the opinions of the International Commission of Zoological Nomenclature or by their botanical counterparts.

The scientific name of the organism may be followed by the name of the describer of the species and the year in which his description was published. When the species is now in a genus different from that used in the original description the describer's name appears in parenthesis. Examples are the name of the cod of the North Atlantic, *Gadus morhua* Linnaeus* 1758, and that of the closely related haddock which is of a different genus, *Melanogrammus aeglefinus* (Linnaeus) 1758. The names are always underlined in typescript and italicized in print. The generic name is capitalized; the specific name is never capitalized (although botanists formerly capitalized specific names derived from proper nouns). In a second use of the name in manuscript, the genus may be abbreviated (as *G. morhua* in the example above).

Unfortunately a correct identification is not always easy or even possible. Species are not fixed, unchanging entities but are stages in the evolution of populations in the generally accepted modern concept. They are not arbitrarily defined entities but are reproductively isolated, interbreeding populations. The evidence of evolution, reproductive isolation, and interbreeding is often scanty however so inferences about species and the practice of identification are based commonly on morphology. Practical and clear definitions of species together with the related definitions of subspecies and genera are given by Milton B. Trautman† (p. 45) which he attributes in part to Ernst Mayr, to which I adhere:

> In bisexual animals a species is a natural population, or a group of populations, normally isolated by ecological, ethological, and/or mechanical barriers from other populations with which they might breed and which usually exhibit a loss of fertility when hybridizing. The above definition applies to all except a few bisexual animals. No single definition can be applicable to all animals however, because in the large series of imperceptible and intermediate stages in the evolutionary transition from a subspecies to a species there is no well-defined gap separating the two; hence it is sometimes impossible to know whether two populations are specifically or only subspecifically distinct.

Dr. Trautman emphasized the fluidity of nomenclature by stressing that the above were his concepts at the time of writing the book. He goes on to comment about the changing of names (p. 46):

> ... It was impossible for the majority of the early describers of fishes such as Rafinesque to review adequately the pertinent literature in order to learn whether their supposedly "new species" had been described, or to compare their specimens with the types of closely related species. As a result, they described many species which had been properly described one or more times previously, and consequently their name for the species,

---

*Carolus Linnaeus is the latinized name of a Swedish botanist, Carl von Linne, who published the tenth edition of "Systema Naturae" in 1758, the initial year of the use of the system of binomial nomenclature.

† Reprinted from "The Fishes of Ohio," by Milton B. Trautman. Copyright 1957 by the Ohio State University Press. By permission of the author and publisher. This quotation is not presented in the original order given by Dr. Trautman.

according to the law of priority, had to be synonymized. Some species were thus unknowingly described many times. . . .

Recent studies of original descriptions of many species have shown that the vague descriptions given by the describers have misled later taxonomists so that they applied scientific names to the wrong species; upon realizing the error another name for these species had to be given. Researchers frequently find a species description in some previously overlooked obscure publication, which description is of an earlier date than the name applied at present to the species. In that case the law of priority demands that the earliest available name is used, even though the other name has been in usage many years. . . .

It was the custom of the earliest taxonomists to place many species in a single genus. . . . Later when the various species were compared with one another certain differences became apparent. As a result, the large genus was divided into many genera, until in some instances the majority of the genera had become monotypic. . . . Still later, when adequate material had been assembled, a monographic study of the group was undertaken, resulting in the accumulation of abundant evidence indicating that a drastic reduction in the number of genera was necessary. . . . This swinging from one genus to many, then back to a few when an adequate study of the group was made, is what normally happens.

. . . The subspecies, or geographic race, is a geographically localized subdivision of a species, which differs genetically and taxonomically from other subdivisions of the species. . . . Unlike species, the subspecies comprising a species can freely interbreed without loss of fertility, but mass interbreeding is usually hindered or prevented by some type or degree of geographical, ecological, ethological and/or mechanical barrier or barriers. As indicated under the species, there are many stages in the evolutionary process of species production from an original homogenous group into two homogenous groups or subspecies. Some subspecies are so similar to one another that the two can be separated only through the application of statistical methods, others are so distinct as to be recognizable at a glance. Taxonomists usually consider a subspecies to be valid if 75% or more of one population can be separated statistically from the other population.

. . . A genus is a systematic unit including one species, or a group of species of presumably common phylogenetic origin, separated by a decided gap from other similar groups. . . . The genus is largely a human conception whose function is to group species in order to stress their relationships to each other. Unfortunately all taxonomists do not place the same value upon the various characters, so they are divided into those who consider any clear-cut character to be of generic value, and those who believe that stressing such characters destroys the purpose of the genus, which is to stress relationship. Such diversity of opinion results in disagreement relative to the number of genera, the "lumpers" stressing relationships, leaving subgenera to stress the minor gaps between groups with a genus, and the "splitters" preferring to give generic rank to what the "lumpers" designate as subgenera.

Organisms are identified in practice from their structure on the basis of their external characteristics as much as possible. The commonly used characteristics of fish are shown in Fig. 3.1. The identification is facilitated by the development of a dichotomous key to the organisms of limited faunal areas (Table 3.1).

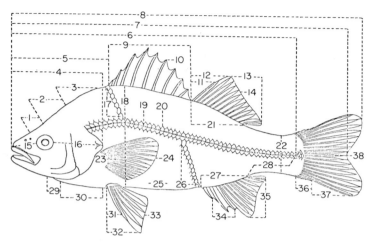

Fig. 3.1. External anatomy of a common spiny-rayed fish illustrating parts and methods of counting and measuring: (1) interorbital, (2) occipital, (3) nape, (4) head length, (5) predorsal length, (6) standard length, (7) fork length, (8) total length, (9) length of base of the spinous or first dorsal fin, (10) one of the spines of the dorsal fin, (11) one of the rays of the second or soft dorsal fin, (12) height of second dorsal fin, (13) length of the distal, outer or free edge of second dorsal fin, (14) one of the soft rays of the second dorsal fin; (15) snout length, (16) postorbital head length, (17) scales above the lateral line or lateral series which are counted, (18) body depth, (19) one of the lateral line pores (since in this figure all of the scales in the lateral series are pored the lateral line is complete), (20) one of the lateral scales which with the remainder form the lateral series, (21) length of base of the second or soft dorsal fin, (22) least depth of the caudal peduncle, (23) the pectoral fin, (24) one of the soft rays of the pectoral fin, (25) abdominal region, (26) scales below the lateral line or lateral series which are counted, (27) length of the base of the anal fin, (28) length of the caudal peduncle, (29) the isthmus, (30) the breast, (31) the pelvic spine, (32) height of pelvic fin, (33) one of the soft rays of the pelvic fin, (34) spines of the anal fin, (35) soft rays of the anal fin, (36) the rudimentary rays, (37) one of the principal rays of the caudal fin, (38) the caudal fin. [Source: Reprinted from "The Fishes of Ohio," Milton B. Trautman. Copyright 1957 by Ohio State University Press, Columbus, Ohio. By permission of the author and the publisher.]

3.2 Classification

After a species has been established it is classified with other species into the larger groups of a hierarchy. Ideally this classification reflects the evolutionary history of the groups; the most closely related are those that have had common ancestors most recently. In practice the evidence for the evolution is usually incomplete or even totally lacking so the relationships must be inferred from the structure. But organisms always resemble each other in some parts, differ in others, so that deciding on the structure that best

TABLE 3.1

Excerpt from a Dichotomous Key for Classification of Fish[a]

Family Salmonidae—Salmons

61	(72)[b]	Rays in anal fin, 8 to 12
62	(69)	Teeth on head and shaft of vomer; spots black (sparse in *S. salar*) (spots may be faint if fish has been in salt water for some time)

Genus *Salmo*—Trouts

63	(66)	No red band along side of body; no red dash below lower jaw
64	(65)	No spots below lateral line
		Atlantic salmon, *Salmo salar*
65	(64)	Large black spot below lateral line, each surrounded by pink or red halo
		Brown trout, *Salmo trutta*
66	(63)	Either red band along side of body or red dash below lower jaw
67	(68)	Red band along side of body; no red dash below lower jaw, no teeth on black of tongue
		Steelhead trout, *Salmo gairdnerii*
68	(67)	No red band along side of body; red dash below lower jaw, teeth present on back of tongue
		Coastal cutthroat trout, *Salmo clarkii clarkii*
69	(62)	Teeth on head of vomer only; spots yellow and red, never black (frequently yellow or red spots may be faint if fish has been in salt water for some time).

Genus *Salvelinus*—Chars

70	(71)	Spots on back, pale yellow; vermiculations on back and dorsal fin absent or weak
		Dolly Varden, *Salvelinus malma*
71	(70)	No spots on back; vermiculations on back and dorsal fin dark green, prominent
		Brook trout, *Salvelinus fontinalis*
72	(61)	Rays in anal fin, 13 to 19

Genus *Oncorhynchus*—Pacific salmons

[a]Source: Clemens and Wilby (1961).

[b]The key is based on the "true-false" method. If statement 61 is false, the reader should look at the number in parentheses, i.e., 72. If the statement is true, the reader should go on to the next statement, i.e., 62.

reflects the evolution is usually controversial. The classification would be much easier if fossils of all intervening forms were available, but of course they never are. The classifier faces the equivalent of a giant jigsaw puzzle with only a small and unknown fraction of the pieces.

As a consequence there is never enough evidence to settle many of the questions about relationships. A classification must have a niche for all forms if it is to be useful, so guesses are made about the relations and separations among groups. Only a few people can hope to master a major group in order to revise the classification on the basis of the extensive and scattered scientific literature. Usually a major revision is a task that occupies the productive lifetime of a man. Nevertheless the disagreements are not important to the fishery scientist who is concerned with a few well-known groups. He must accept the differences until better evidence accumulates.

Many modern fish classifiers follow Berg's (1940) classification of fishes for the following reasons: (1) it is recent enough to include most of the evidence available, (2) it is comprehensive including all known living fossil forms, (3) it tends to show relationships by lumping groups instead of splitting them, and (4) it uses a reasonably standard system of units and endings. A more recent popular classification is in Greenwood *et al.* (1966). The student will find, however, several other classifications in common use; there are almost as many variations as the number of senior taxonomic ichthyologists. In this book the general classification of fish outlined by Lagler, Bardách, and Miller (1962) and for North American fish the list recommended by the American Fisheries Society in 1960 are followed, both of which are close to the classification set forth by Berg.

The standard units of the hierarchy are species, genus, family, order, class, and phylum to each of which the prefixes *super-* or *sub-* may be added (Table 3.2). Endings of *-idae* for animal families and *-inae* for subfamilies are in almost universal use, but endings for other units are more variable. Berg uses a common system of endings of *-oidae* for superfamily, *-oidei* for suborder, and *-iformes* for order. The ending of *-aceae* is commonly used for plant families.

3.3 Major Groups

What are the major kinds of aquatic organisms? The student of botany or zoology will easily find complete discussions of the classification of plants and animals so it is unnecessary to summarize them here. Instead it seems desirable to discuss the classifications and general habits of the most useful groups because many of these common animals are less important taxonomically than the bizarre forms which may have no use at all. The groups of plants and animals that are involved primarily in the aquatic production

TABLE 3.2
SCIENTIFIC CLASSIFICATION OR COASTAL CUTTHROAT TROUT

Group	Name	Meaning
Phylum	Chordata (Vertebrata)	Having notochord
Subphylum	Craniata	Having cranium
Superclass	Gnathostomata	Having jaw mouth
Class	Teleostomi	Bony fish
Subclass	Actinopterygii	Rayed fins
Order	Clupeiformes	Herringlike
Suborder	Salmonoidei	Salmonlike
Family	Salmonidae	Salmonlike
Subfamily	Salmoninae	Salmonlike
Genus	*Salmo*	Salmon
Species	*clarkii*	(Names in honor
Subspecies	*clarkii*	of a Mr. Clark)

Scientific name:

Salmo clarkii clarkii Richardson 1836
    Genus—always capitalized
    Species and subspecies—never capitalized
    Name always in two or three parts which are underlined
      in typescript, italicized in print
    Name and date identify original species describer
    Name in parentheses if species now in different genus

cycle and the principal groups that are harvested by man have been chosen. The latter are those that yielded more than about 100,000 tons (about 0.2% of total world production) for human use in 1967 according to FAO statistics (Table 3.3). Many of these useful groups fit the commonly accepted family units of organisms but others embrace the more inclusive units of the heirarchy. Family or larger groups in the following discussion have been chosen somewhat arbitrarily.

3.3.1 PHYTOPLANKTON

Plankton, which is commonly defined as the animals and plants that drift in either fresh or salt water, is classified according to size and ability to conduct photosynthesis: *phytoplankton* when it can and *zooplankton* when it cannot. Nonliving material such as inorganic silt or organic detritus is excluded. Plankton ranges in size from the smallest of living forms that can be recognized, such as bacteria close to 0.001 mm in diameter, to jellyfish several meters long. *Nannoplankton* consists of the plants and animals that pass through nets of the finest mesh that are practical to use (with openings of about 0.05 mm in diameter). These organisms are separated from the

TABLE 3.3
INTERNATIONAL STANDARD STATISTICAL CLASSIFICATION OF AQUATIC ANIMALS AND
PLANTS[a]

Division	Percentage of world catch by weight in 1967 (whales excluded[b])
Division 1 Freshwater and Diadromous Fishes	13.6
Group 11 Freshwater fishes	10.8
12 Sturgeons, paddlefishes	Ø[c]
13 River eels	0.1
14 Salmons, trouts, smelts, etc.	1.8
15 Shads, milkfishes, etc.	0.9
Division 2 Marine Fishes	77.6
Group 21 Flounders, halibuts, soles, etc.	2.0
22 Cods, hakes, haddocks, etc.	13.5
23 Redfishes, basses, congers, etc.	5.2
24 Jacks, mullets, etc.	3.4
25 Herrings, sardines, anchovies, etc.	32.5
26 Tunas, bonitos, skipjacks	2.2
27 Mackerels, billfishes, cutlassfishes, etc.	4.4
28 Sharks, rays, chimaeras	0.7
29 Unsorted and unidentified fishes	13.7
Division 3 Crustaceans, Mollusks and other Invertebrates	7.4
Group 31 Crustaceans	2.2
32 Mollusks	5.1
33 Sea cucumbers, sea urchins, ascidians, etc.	0.1
Division 4 Whales	(60[b])
Group 41 Blue whales, fin whales, sperm whales, etc.	(52)
42 Minke whales, pilot whales, etc.	(8)
Division 5 Seals and Miscellaneous Aquatic Mammals	Ø
Group 51 Porpoises, dolphins, etc.	
52 Eared seals, hair seals, walruses, etc.	
53 Miscellaneous aquatic mammals	
Division 6 Miscellaneous Aquatic Animals and Residues	0.1
Group 61 Turtles, frogs, etc.	0.1
62 Pearls, shells, sponges, corals, etc.	Ø
Division 7 Aquatic Plants	1.3
Group 71 Aquatic plants	1.3

[a] From "FAO Yearbook of Fishery Statistics," Vol. 24. 1968.
[b] Whale catches are not reported by weight. The figures shown are thousands of animals.
[c] Ø = < 0.5%.

water by filtration through membrane filters in the laboratory or by centrifugation. The intermediate-sized plankton, called either *microplankton* or *net plankton*, is retained by plankton nets. Most of the larger *phytoplankton* is retained by nets with openings of about 0.05 mm in diameter but passes through openings of about 0.2 mm in diameter. The larger net plankton, including most adult crustacean zooplankton, is retained by nets with openings of about 0.2 mm in diameter. The largest plankton or macroplankton includes animals such as jellyfish and the intermediate-sized crustacea such as euphausids which may be several centimeters long. It will be obvious that there is no precise point of separation either between the size classes of plankton or between drifting *plankton* and the swimming *nekton* which includes the fish; both are part of the *pelagic* community of organisms.

Although nannoplankton is extremely difficult to collect and study, it appears that the relative masses of the different groups are inversely proportional to the sizes of individuals. The nannoplankton comprises the great mass of living material in most waters, the microplankton a lesser mass, and the macroplankton the least.

The phytoplankton converts the energy of the sun to organic material and by doing so supports almost the entire web of life in the sea. It can do this not by virtue of size as most of the terrestrial plants do but by virtue of number; vast numbers are spread throughout the photosynthetic zone of all waters. The bottom of the photosynthetic zone is the level reached by about 1% of the sunlight; it is about 150 m in the clearest water, 10 m in typical coastal waters, and less than 1 m in turbid waters. Phytoplankton falls out of the photosynthetic zone and can live in darkness for a time but is rarely found below 200 m.

The phytoplankton is predominantly composed of diatoms (Fig. 3.2) of the class Bacillariophyceae. These are one-celled algae that have a cell wall of silica that is more or less covered by a jelly. Some live individually; some adhere together in colonies or filaments. They usually reproduce by simple cell division, occasionally by a kind of sexual reproduction. Their storage products from photosynthesis are predominantly fats and oils in marked contrast to higher plants which store carbohydrates. Not all diatoms are planktonic; some live on the bottom in shallow water where they may form a thick slime.

Most of the rest of the phytoplankton are flagellated organisms that can be considered either plants or animals. They swim by means of either one or two whiplike flagella but possess either chlorophyll or other similar pigment that enables them to feed as plants. The principal group is that of the *Dinoflagellates* which have two flagella and a prominent groove around the body (Fig. 3.2).

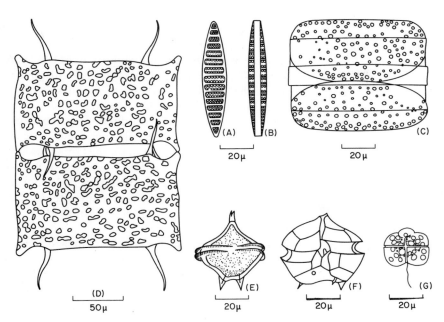

Fig. 3.2. Some common phytoplankton. (A,B) Top and side views of *Fragilariopsis*, a diatom that is the principal food of the krill of the southern oceans, (C) *Coscinodiscus*, a diatom, (D) *Biddulphia* a diatom, (E) *Peridinium*, a dinoflagellate, (F) *Gonyaulax*, a dinoflagellate frequently involved in paralytic shellfish poisoning, (G) *Gymnodinium*, a dinoflagellate that occasionally causes "red tides."

A characteristic feature of phytoplankton populations is the recurrence of "blooms." When environmental conditions improve as with spring warming in temperate latitudes or an increase in nutrients, phytoplankton reproduce very rapidly. One or a few species find some advantage and predominate in the population. They may become abundant enough to color the water or even to make it resemble a thin soup. After gaining predominance one species commonly gives way to another and that one in turn to another so that there is a succession of blooms through the growing season. Such blooms of plant material may radically increase the food for the grazing animals, and these in their turn may flourish and provide increased food for the carnivores. Other blooms of plant material may have an opposite effect and condition the water in way injurious to animals. An example is a bloom of dinoflagellates called a "red tide," which may kill large quantities of fish or even be directly irritating to people who come in contact with the water.

3.3.2 ZOOPLANKTON

The zooplankton includes a vast assortment of animals from tiny, one-celled protozoans which may be part of the nannoplankton to crustaceans and jellyfish. It includes a large group of animals that is permanently planktonic as well as a large group of the larval forms of worms, mollusks, crustaceans, and fish that become nektonic or benthic later in life.

Zooplankton is of special interest in the aquatic production cycle because it includes a large proportion of the grazing animals that feed on plant material. These are the larval and older forms that are either small enough to subsist by eating individual phytoplankton cells or possessed of a filter apparatus that will retain large numbers of phytoplankton. Thus not all plant-eating (i.e., grazing) animals are filter feeders, and many filter feeders do not eat plants, they feed on tiny animals through a filter system that is not fine enough to retain phytoplankton.

The predominant grazing animals of the water are the Crustacea which swarm in the sea as insects swarm on land. Many thousand kinds exist, but the most important group is the subclass Copepoda. The copepods are mostly of pinhead size though individuals of one of the more abundant genera, *Calanus* (Fig. 3.3), are about 3 mm long. Like other Crustacea, the copepods grow by moulting. The egg is either free in the water or carried by the female and it hatches into a larva that goes through several moults before maturing. The copepods of the Arctic apparently breed only once a year, whereas the tropical species may have several generations.

Another prominent group of grazing zooplankton consists of the tiny shrimplike Crustacea of the high seas, the euphausids. Most of them are less

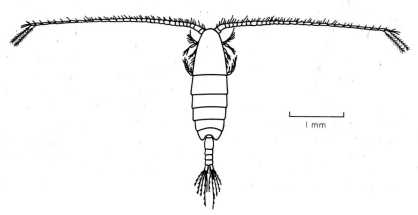

1 mm

Fig. 3.3. *Calanus finmarchius*, one of the common copepods of the northern oceans.

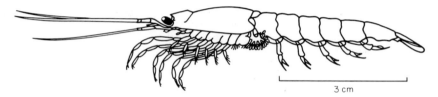

3 cm

Fig. 3.4.   *Euphausia superba*, the krill of the southern oceans that is the major food of baleen whales.

than 1 cm long, but the group includes the krill (Fig. 3.4) ranging from 2 to 7 cm long which comprise the primary food of the baleen whales of the Arctic and Antarctic Oceans.

3.3.3   RHODOPHYTA AND PHAEOPHYTA—RED AND BROWN ALGAE

The aquatic seed-producing or higher plants live along the sheltered edges of the waters at depths of less than about 3 m. These include rushes, sedges, grasses, and lilies many of which have been useful to man through the centuries. These remain useful to small groups of people in many parts of the world, but some of them become serious nuisances at times (e.g., water hyacinth). They are not, however, either a major part of the biomass of aquatic plants or a major part of the commercially important plants.

The most commercially valuable marine plants are the red and brown algae which are commonly known as *seaweeds*.* The seaweeds are attached to the bottom in depths to about 20 m by a holdfast that superficially resembles a root of a higher plant. Above the holdfast they have a stemlike stipe that connects to filaments or blades, in which most of the photo synthesis occurs. The whole plant may be more than 100 m long in some species, such as the giant kelp *Macrocystis* (Fig. 3.5) of the Northeast Pacific. Reproduction of these algae is highly varied but commonly involves a sexual generation that produces gametes and an alternate generation that produces spores. Either or both generations may be conspicuous; in the brown algae the sporophytes are large, the gametophytes tiny. Both spores and gametes are microscopic and either or both drift (or swim actively) with the water currents.

The red and brown algae are used directly for human and stock food in many countries; this use is extensive, however, only in the Orient. Here many kinds are eaten regularly. Nori, the brown algae of the genus *Porphyra*,

---

*Some of the other red and brown algae are unicellular and some of the other red algae form parts of coral reefs. In fresh water another group, the green algae, predominate.

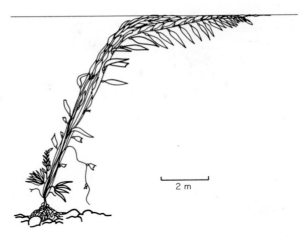

2 m

Fig. 3.5. A mature plant of the giant kelp *Macrocystis pyrifera*, a brown algae of the Pacific Coast of North America.

is cultivated extensively in Japan. In other parts of the world the seaweeds are valuable principally for their colloids which are extracted. The brown algae produce algin, and the red algae agar and carrageenin; all are thickening agents useful in the formulation of drugs, paints, and foods.

3.3.4 DECAPODA—SHRIMPS, CRABS, AND LOBSTERS

The many thousands of species of Crustacea were separated by the earlier naturalists into the Entomostraca, the tiny insectlike forms, and the Malacostraca, the larger forms. This division suits our needs here (even though the Entomostraca have since been divided into several groups) because the Entomostaca are of little direct use to man. Most Malacostraca are bottom-living when juveniles or adults and include as a principal group the Decapoda (ten feet), which are taken by the fisheries. These are crustaceans with compound eyes and a carapace that is fused to the thoracic segments. The shrimps and prawns are decapods with long abdomens and with body compressed from side to side. The lobsters and crayfish are larger decapods with long abdomens and with body compressed from top to bottom. The edible crabs are decapods with abdomens usually short and folded beneath the body.

Like all crustaceans, the decapods grow by moulting. The moults occur in quick succession during larval stages, when the animals grow rapidly. The frequency of moulting decreases as the animals mature to a common rate of once a year in the larger crabs and lobsters, and perhaps only once in several years in very old animals. In many mature female Crustacea a moult immediately precedes mating during which sperm are deposited in the female to be used later to fertilize the eggs as they are laid. The newly fertilized eggs of crabs and lobsters are commonly attached to abdominal appendages and carried by the female until hatching (eggs of some shrimp are discharged into the water). After hatching the larvae of most Crustacea drift or swim with the ocean currents for a time and then descend to the bottom where they live as juvelines and adults. In crab and lobster fisheries conducted by pots commonly the egg-carrying females are protected by law and must be released alive.

Many species of shrimps and prawns (the latter word frequently means merely large shrimps) are sought by the fisheries, but two groups are sought above all others. Most important are shrimps of tropical and subtropical waters around the world belonging to the family Penaeidae (Fig. 3.6). The valuable species of these shrimps spawn regularly in the open sea; the larvae pass through several developmental stages while drifting toward land and then grow rapidly for a time as juveniles in estuaries. They then return to the sea bottom to mature in depths of 10–150 m where they are caught by trawls. The whole life cycle is usually completed in about 1 year. The fisheries for these shrimps are usually bottom fisheries on the continental shelves near or in estuaries.

The other major group of shrimps are the cold-water shrimps of the family Pandalidae. These are found on the continental shelves to 200 m and in the deep estauries of the temperate and arctic waters of the Northern Hemisphere. The most common species is *Pandalus borealis* of both the Atlantic and Pacific Oceans. The shrimps of this family mature first as males and later become females during a lifespan of between 3 and 5 years. In consequence the mature females are preferred by the fisheries, and some stocks have been

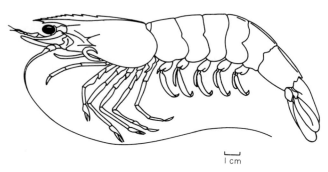

I cm

Fig. 3.6. The brown shrimp, *Penaeus aztecus*, family Penaeidae, one of the principal commercial shrimps of the Caribbean.

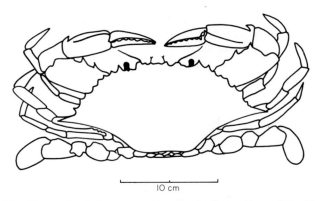

10 cm

Fig. 3.7.   The blue crab, *Callinectes sapidus*, family Portunidae, of the North Atlantic Ocean.

depleted quickly. Other stocks seem to be very large such as some in the Bering Sea.

The crab fisheries seek only a few of the thousands of species of crabs, most of which are larger species in three families. The Cancridae includes *Cancer*, a genus of edible crabs of the Northeast Atlantic and North Pacific (where one species is the Dungeness crab of western North America). The Portunidae, a family of swimming crabs, includes the genera *Portunus* and *Callinectes* of Europe and North America which are commonly known as "blue crabs" (Fig. 3.7). The Lithodidae is a family of long-legged crabs in

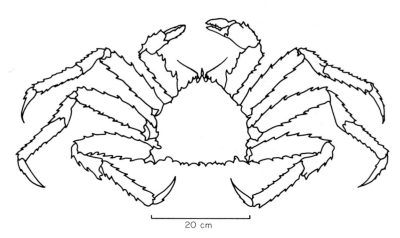

20 cm

Fig. 3.8.   The king crab *Paralithodes camtschatica*, of the North Pacific Ocean and Bering Sea.

which the posterior pair of legs is degenerate and hence appear to be 8-legged crabs. This family includes the genus *Paralithodes*, a species of which is the king crab of the North Pacific and Bering Sea (Fig. 3.8). Another member of the family is the centolla of the continental shelves near southernmost South America.

The lobsters belong to two families of large marine Crustacea that live on the continental shelves. The Homaridae (Fig. 3.9) includes the large (to 13 kg) lobsters of the North Atlantic genus *Homarus* which have very large pinching claws on the first pair of legs. The Norway lobster, *Nephrops*, of the Northeast Atlantic and the freshwater crayfish are closely related. The family Palinuridae includes the large sea crayfish or spiny lobsters (Fig. 3.10) of tropical and subtropical waters around the world. These lobsters have no pinching claws on the legs.

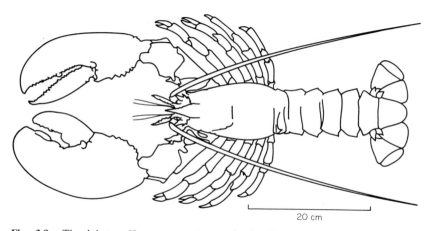

20 cm

Fig. 3.9. The lobster, *Homarus americanus*, family Homaridae, of the western North Atlantic Ocean. It is very similar to *H. gammarus* of European waters.

3.3.5 LAMELLIBRANCHIATA—BIVALVE MOLLUSKS

These are mollusks with two shells (two valves), a mouth, but no head that are bilaterally symmetrical or nearly so. They are all aquatic and mostly marine animals that live on the bottom in the photosynthetic zone. Most of them maintain currents of water through their bodies by means of cilia and feed by separating the microscopic plants and animals from the water current. About 11,000 species have been described of which only a very few are eaten by man.

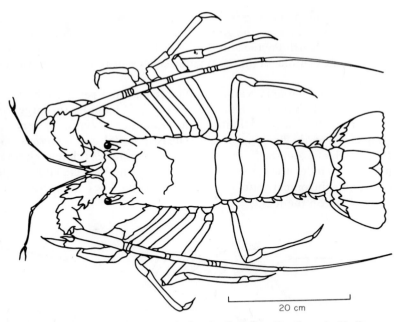

Fig. 3.10.   The spiny lobster, *Palinurus elaphas*, family Palinuridae, from the Mediterranean.

The most diverse group of commercially important species is that of the hard-shell clams which belong to several families. The taxonomy of these is too complicated to present here, but in general they are symmetrical bivalves with a foot and with two calcareous shells that close tightly together. They burrow in the bottom and must be harvested by digging or dredging. Included are the genera *Venus* and *Cardium* (cockle) of Europe, *Mercenaria* (quahaug) (Fig. 3.11), *Spisula* (surf clam), *Arctica* (ocean quahaug), *Saxidomus* (butter clam) of North America, *Venerupis* and *Meretrix* of Eastern Asia and many others. Most of these are harvested from wild stocks but some are cultivated, especially *Venerupis* in Japan.

Another group of clams are the soft-shell clams of the family Myacidae. They have a gaping, fragile shell. *Mya* (Fig. 3.12), the principal North American genus, lives primarily in mud flats of the intertidal zone on both Atlantic and Pacific coasts.

The mussels of the family Mytilidae are bivalves that have shells covered with a noncalcareous outer layer and attachments called "byssi." They occur in large beds. The important genera are *Mytilus* of Europe and *Aulacomya* of South America. *Mytilus* (Fig. 3.13) is abundant in North America but is little in demand perhaps because of the occurrence of paralytic shellfish

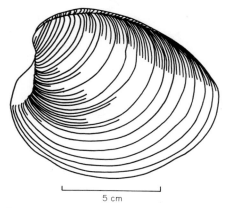

Fig. 3.11. The northern quahaug, *Mercenaria mercenaria*, family Veneridae, of the Atlantic Coast of North America.

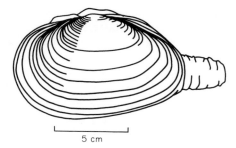

Fig. 3.12. The soft shell clam, *Mya arenaria*, family Myacidae, of the Atlantic and Pacific Coasts of North America.

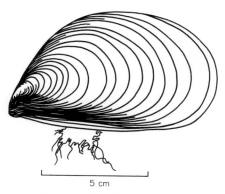

Fig. 3.13. The bay mussel, *Mytilis edulis*, family Mytilidae, is cultivated in Europe and grows naturally from Arctic to subtropical waters on both coasts of North America and Northeast Asia.

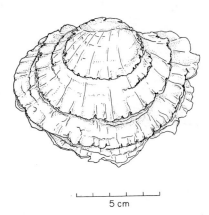

Fig. 3.14. The European oyster, *Ostrea edulis*, family Ostreidae.

toxins in the species of the Pacific coast. Mussels are cultivated extensively in Europe and to a lesser extent in Eastern Asia.

The Ostreidae includes the two important genera of edible oysters, *Ostrea* and *Crassostrea*, which are cultivated extensively in several parts of the world. The family is distinguished by the single adductor muscle, dissimilar shell valves, attachment of the left shell, absence in the adult of a foot, absence of a byssus, and triangular ligament (Fig. 3.14).

Ostrea is the flat oyster with the left valve not deeply cupped and more or less circular; *Crassostrea* is the larger, elongated oyster with the left valve deeply cupped. The genera have been separated also becuase of the structure

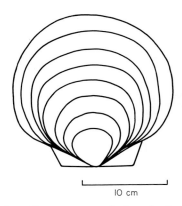

Fig. 3.15. The giant scallop, *Placopecten magellanicus*, family Pectinidae, of the western North Atlantic.

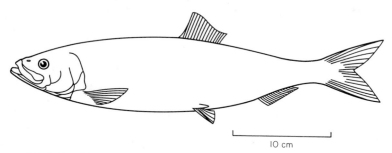

10 cm

Fig. 3.19.   The North Atlantic herring, *Clupea harengus*, family Clupeidae.

Arctic Ocean and south to California. The genus is divided into a number of species and subspecies which are not easily distinguishable, but more importantly it is comprised of dozens of different races, each with distinctive life habits. Some migrate thousands of kilometers annually, others remain in the same bay. Most races spawn in spring, others spawn in summer or fall. Some grow rapidly and rarely exceed 5 years of age, others may live to be 25 years old. Some races may spawn in the intertidal zone, others in 200 m of water. The eggs are adhesive and laid on the bottom. In some small races the females produce less than 10,000 eggs each at every spawning, in others more than 100,000. The sea herrings eat zooplankton with an extraordinary efficiency. They are found where the zooplankton is abundant whether the plankton is composed predominantly of copepods, euphausids, crab larvae, molluskan larvae, worm larvae, or fish larvae.

In most parts of the world herring are either salted, smoked, or dried. Many small herring are canned and called *sardines* and as such should not be confused with the genera *Sardinops*, *Sardinia*, or *Sardinella* which are properly called *sardines*. Considerable quantities are processed for the oil and dried to fish meal. The oil is a valuable base for paint and ink; the meal is used as a vital ingredient of domestic animal and poultry rations.

The genus *Sprattus*, the sprat, a small herring of the Northeast Atlantic, Mediterranean, and Black Seas, is similar to *Clupea* in habit and forms the basis for an extensive fishery where it occurs. The genus *Clupeonella*, also a sprat, supports a valuable fishery in the Sea of Azov, Black Sea, and Caspian Sea.

The genera *Alosa*, the shad of Europe and North America, *Ilisha*, the shad of India, and part of the species of *Caspialosa*, the shad of the Black and Caspian Seas, are larger herrings that migrate up rivers to spawn. One North American species, *Alosa pseudoharengus*, the alewife, usually enters fresh water only to spawn but may become landlocked. Recently this species has become the predominant fish in the Great Lakes of North America and apparently could support a large fishery.

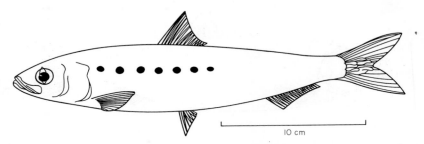

Fig. 3.20. The Japanese pilchard (Ma-iwashi), *Sardinops melanosticta*, family Clupeidae, one of the important commercial fishes of Japan.

Sardinops, the pilchard or sardine (Fig. 3.20), is the second most important genus of the family. Major stocks occur off Japan, Western United States, Southwest Africa, Australia, New Zealand, and Western South America. It is a small, rapidly growing herring, the usual commercial sizes being between 10 and 25 cm and ages 1–7 years. It may reach 30 cm and 15 years, however. *Sardinops* are found primarily in cool, upwelling waters of temperate latitudes which are rich in plankton. They feed on either phytoplankton or zooplankton or both by filtering the smaller forms out of the water and striking at the larger. Spawning may occur throughout the year or in a certain season in some races. Most stocks migrate a few hundred kilometers each year.

Sardinia, the sardine of the Mediterranean regions, is similar to *Sardinops* in habits. It is a small herring; those caught by the commercial fishery are usually between 10 and 20 cm long and less than 3 years old. It spawns in autumn and winter. It is nowhere as abundant as species of *Clupea* or *Sardinops* because the Mediterranean waters are relatively barren, but it supports important local fisheries.

Sardinella is the major genus of tropical sardines. It also is a small herring, rarely reaching more than 20 cm long and more than 3 years old. The principal stocks occur in the Arabian Sea, off Southeast Asia, in the Gulf of Guinea, and off Northwest Africa, but one species or another can be found in most tropical coastal waters of the world. It supports many primitive fishermen who commonly attract the fish to the surface with lights and catch them with either dipnets or enveloping nets.

Brevoortia, the menhaden, occurs off Eastern North America from the Gulf of Mexico to the Gulf of Maine and along Eastern South America off Southern Brazil and Northern Argentina. It is a medium-sized herring measuring on the average between 30 and 40 cm in length, and lives up to

10 years. Unlike most other herrings it feeds almost entirely on diatoms and the smaller zooplankton which it obtains through a filtering system that is one of the more efficient known among fish. It has been reported that an adult can filter 25 liters/minute.

The menhaden is too bony to provide good food and is useful to the fisheries for oil and fish meal. In recent years the landings of menhaden in the United States have been greater in volume than those of any other species taken by United States fishermen.

General.   When one compares the world distributions of all the herrings (including those minor genera that were not discussed) and the anchovies (to be discussed next) with the biologically enriched areas of the oceans, it appears that almost every enriched part of the ocean has an abundance of one species from one family or the other. *Clupea* and *Sprattus* occur in the cold waters of the north (not the south), *Sardinops, Brevoortia*, and *Engraulis* in the temperate and subtropical waters; and *Sardinella* in the tropical waters. These fish are the principal converters of energy from phytoplankton and zooplankton to fish flesh; they are the predominant small pelagic fish of the world.

It is interesting that few are known to occur in abundance in the enriched areas of the southern oceans, but this dearth may be due to a lack of exploration. It is interesting also that rarely are two species from these families equally abundant in the same area. When two species occur along the same coast, they do so either at different depths or at different seasons; one species predominates and seemingly excludes the other, but this problem is one of population ecology and it will be discussed in a later chapter.

3.3.9 ENGRAULIDAE—ANCHOVIES

The total production of the anchovy family in 1967 was 11.6 million tons, or about 19% of the total world fishery production. More than 90% of this catch was taken off Peru where the fish are cooked, separated from their oil, and dried into meal for animal food. The balance is taken at scattered places off Northwest Europe, Japan, Australia, and the Mediterranean. Part of this catch was salted and used directly for human food, the balance converted to oil and meal. Unexploited stocks exist; on the basis of egg and larval surveys the standing crop of California anchovy is known to amount to close to five million tons and practically none of it is being utilized.

The anchovy family includes only about 6 genera and 40 species. These are widely distributed in tropical, subtropical, and temperate waters. They resemble herring superficially but have a larger mouth and a rounded belly. They are small fish; few exceed 20 cm in length; most live 3 years or less.

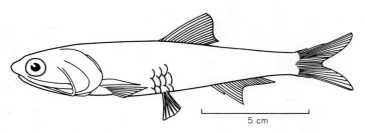

Fig. 3.21.   The Peruvian anchovy, *Engraulis ringens*, family Engraulidae, the species that supports the largest fishery in the world.

Like the herrings they feed predominantly on the tiny animals that comprise the bulk of the zooplankton such as the copepods, larval crabs, and larval mollusks.

Almost all of the commercial anchovies of the world belong to the genus *Engraulis* (Fig. 3.21). The major stock is found off Peru and North Chile; lesser stocks are found in the western North Pacific, off Australia, in the Mediterranean and Black Seas, and in the Atlantic off Europe.

3.3.10 SALMONIDAE—SALMON AND TROUTS

The Salmonidae includes some of the most exciting fish of the world either to eat or to catch for sport. Their flesh is usually richly flavored and commands a premium price. Some offer a spectacular display when hooked on angling gear and anglers may pay hundreds of dollars per day for the privilege of seeking them. Most species inhabit clean, well-oxygenated, colder waters of the Northern Hemisphere. Several species have been introduced into suitable waters in the Southern Hemisphere.

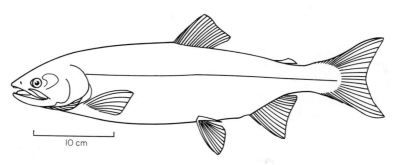

Fig. 3.22.   The pink salmon, *Oncorhynchus gorbuscha*, family Salmonidae, the most abundant of the Pacific salmons.

The family includes three important genera. *Salmo* contains the Atlantic salmon, the rainbow trout, brown trout, and steelhead trout. *Oncorhynchus* (Fig. 3.22) includes the six species of Pacific salmon, one of which occurs only in Eastern Asia, but the rest of which are found around the North Pacific rim from Japan to California. *Salvelinus*, the genus of the chars, has the brook trout, lake trout, Dolly Varden trout, and Arctic char. Another less widely known genus is *Hucho*, a large predatory trout (to 1.5 m) of the Danube River. The family is characterized by the presence of an adipose fin, fins without spines, and smooth (cycloid), small scales. Closely related fishes are the whitefish of the family Coregonidae (sometimes named in the Salmonidae as a subfamily) and the smelts of the family Osmeridae.

Many members of the Salmonidae family have an unusual tolerance for either fresh water or salt water. The Atlantic salmon, Pacific salmon, steelhead trout, and Dolly Varden trout are commonly anadromous; they spawn in fresh water, move to sea in the juvenile stage to feed, fatten, and mature, and return to fresh water to spawn. Some of the other species such as brown trout and brook trout may go to sea occasionally. On the other hand, all of the anadromous species can spend their lives in fresh water, although usually they do not grow as rapidly or reach as large a size as they do in the sea. A recent spectacular example of fresh water life for an anadromous fish has been the successful introduction of the coho salmon, *Oncorhynchus kisutch*, into the Great Lakes of North America.

The anadromous habit of many species and the ability of individuals to return to their home stream in which they were born has attracted much attention. Some species undergo spectacular color changes as they prepare to spawn. Individuals of some species make regular migrations for thousands of kilometers at sea; indeed the Atlantic salmon from Norwegian home streams may feed off Greenland, and the Pacific salmon from Japanese home streams may feed in the Gulf of Alaska.

All of the members of this family are carnivorous. When young their diet consists usually of either small crustaceans or insects in fresh water and small crustaceans in the sea; as they become older commonly their diet includes a greater proportion of small fish.

Most of the members of the Salmonidae produce relatively few (less than 5000) large eggs. Commonly the females lay their eggs in a nest or redd that they build in gravel through which water is circulating. The males fertilize the eggs and then the females cover them with gravel. The eggs hatch after an incubation period of between 30 and 150 days into relatively large, agile larvae. The larvae of several species, notably the rainbow trout, will accept artificial food as soon as they begin to feed; therefore such species enjoy a major advantage in aquaculture.

3.3.11 CYPRINIDAE—CARPS AND MINNOWS

The Cypriniformes are unique in having a number of small bones that connect the swim bladder to the hearing apparatus in the skull. Most of them are freshwater fish; some occur on all of the continents. They include the common characins of Africa and America in the family Characidae, many families of catfishes of the suborder Siluroidei (see Section 3.3.12), and the carps and minnows of the family Cyprinidae.

The Cyprinidae contains more species than any other family of fish. They are freshwater fish of Africa, Asia, North America, and Europe but unless artificially reared they are not found in Australia or South America. They possess no teeth on the jaws but normally have well-developed-teeth on the pharyngeal bones. Their fins are usually without spines. Usually they possess a single dorsal fin. Ordinarily the bodies are covered with smooth cycloid scales, though occasionally they may be naked.

The biology of the cyprinids is highly variable. A few live in cold waters and require a large supply of oxygen, but many prefer warm waters and some tolerate an amazingly low quantity of oxygen, as little as 0.5 mg/liter. Most female members of the family deposit large numbers of eggs; these are fertilized externally and usually are not cared for. Most cyprinids do not make long migrations from feeding areas to spawning grounds. Larval cyprinids usually feed on zooplankton but the older animals eat many kinds of food. Most juveniles and adults feed on bottom animals. Some adults live on other species of fish; some are herbivores and feed either on phytoplankton or detritus and a few even on higher plants.

Hundreds of species of the Cyprinidae are small fish. These include the minnows, daces, chubs, and shiners. These feed predominantly on insects and zooplankton throughout their lives and in turn provide forage for larger predatory fish.

A few species of the family provide highly important food fishes in Europe, Asia, and Africa. These include the roaches, the breams, the carps, and the tench. The wild carp of the genus *Cyprinus* (Fig. 3.23) lives in many of the sluggish rivers and lakes of Europe and Asia where it is subject to intensive fishing. The species has been bred for thousands of years and several domesticated varieties have been developed (see Section 7.3). Other genera of Asian carps of the family Cyprinidae are sought likewise in the wild and grown in captivity; these include *Catla, Labeo, Barbus*, and *Cirrhina.*

Another important genus is the crucian carp, *Carassius.* This fish is closely related to the wild carp. It is extremely hardy, grows to a length of 45 cm, and serves as a food fish in parts of Europe and Asia. A form of *Carassius*, the goldfish, has been selectively bred into hundreds of varieties for aquariists.

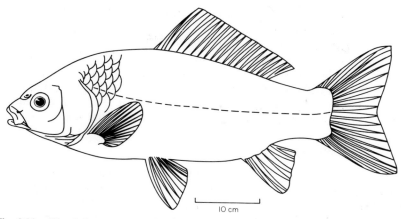

Fig. 3.23. The Asian carp, *Cyprinus carpio*, family Cyprinidae, a species introduced over much of Europe and North America and extensively cultured in many parts of the world.

3.3.12 SILUROIDEI—CATFISHES

The Cypriniformes includes the suborder Siluroidei to which belong about 28 families of catfishes. Most of them lack scales but some have spines or plates. Usually they have a head with several pairs of barbels and a body flattened from top to bottom. Many have a single strong spine in some of the fins.

The catfishes are of special importance as food fish because they have few bones in the flesh and most are especially tolerant of poor water with small quantities of oxygen. Some, such as the walking catfish, can even live on land in moist vegetation for hours at a time. Most of the species build nests; commonly the males guard the eggs and young. Most are predatory

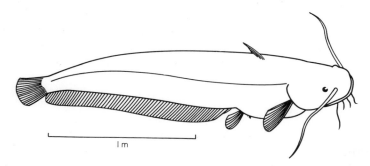

Fig. 3.24. The European catfish, *Silurus glanis*, family Siluridae, one of the largest fresh-water fishes.

bottom feeders on invertebrate animals or other fish. Some species are well adapted to freshwater aquaculture.

The suborder includes the Old World catfishes of the family Siluridae (Fig. 3.24) which may reach 2.5 m in length and 300 kg in weight. Others are the Bagridae of Africa and Asia, the Clariidae of Africa and Southeast Asia, and the Ictaluridae of North America. Several diverse families occur in South America some of which are spiny or armored. Two families, Bagridae and Plotosidae, include species that live in the sea.

3.3.13 SCOMBERESOCIDAE—SAURIES

The sauries are little-known, small (< 45 cm), pelagic fishes of tropical, subtropical, and temperate parts of the open ocean. Their closest relatives are the half-beaks, family Hemiramphidae; flying fishes, family Exocoetidae; and needle fishes, family Belonidae. They spawn in the open sea. Their eggs and larvae are pelagic. Their food consists of zooplankton, small crustaceans, and small fishes. The principal commercial fisheries are in the subtropical and temperate waters of the Northern Hemisphere.

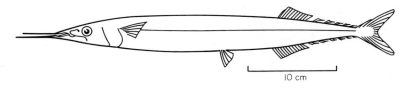

Fig. 3.25. The North Atlantic saury, *Scomberesox saurus*, family Scomberesocidae.

The Atlantic saury, *Scomberesox* (Fig. 3.25), is distributed widely in the Atlantic as far north as Newfoundland and northern Norway. It occurs also in the Mediterranean and Black Seas. The Pacific saury, *Cololabis*, inhabits the waters of both the Asiatic and American coasts, north to Sakhalin and Alaska. The flesh of these sauries is moderately oily, very tasty, and in great demand when available. They are taken with floating gill nets or purse seines but most successfully at night by dipnet after having been attracted to the surface by lights.

3.3.14 GADIDAE—CODS AND HAKES

The gadoids, or Codlike fishes, include only about 70 species, but they are the second most important family of food fishes in the world (the first is Clupeidae). The world production in 1967 was about 8.1 million tons—a record production but an amount that reflects the long, slow growth of the

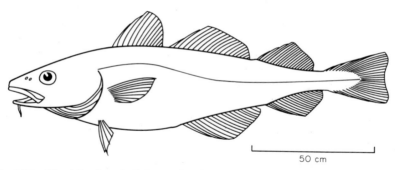

50 cm

Fig. 3.26.   The Atlantic cod, *Gadus morhua*, family Gadidae, a species of major importance to many countries.

fisheries from prehistoric times. The gadoid fish populations tend to be relatively stable in production.

The Gadidae are characterized by lacking spines in the fins, by having ventral fins located far foward on the belly, and by being physoclistous, i.e., when an air bladder is present it has no connection with the gut. Most of the family have either three dorsal fins and two anal fins or second dorsal and anal fins with very long bases. They also lack intramuscular bones, a feature that makes them easy to eat.

The family is usually divided into three subfamilies. The Gadinae includes *Gadus*, the cod (Fig. 3.26); *Odontogadus*, the merlang of Europe; *Melanogrammus*, the haddock; *Pollachius*, the pollock or saithe; and *Theragra*, the walleye pollock. The Merlucciinae contains only the genus *Merluccius*, the hake (Fig. 3.27). The Lotinae, much less commercially important, has *Brosme*, the cusk; *Urophycis*, another hake; and *Lota*, the single freshwater genus.

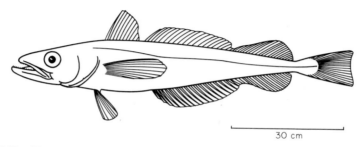

30 cm

Fig. 3.27.   The Pacific hake, *Merluccius productus*, family Gadidae, an abundant species off the west coast of North America.

The commercially important genera are all marine. All live in temperate or polar waters predominantly in the Northern Hemisphere with the exception of *Merluccius* which is abundant off South Africa, Argentina, and Chile, as well as off Europe, and both coasts of North America. All live and feed on or near the bottom of the continental shelf or continental slope except *Merluccius* and *Theragra* which live at times near the surface or at intermediate levels. All migrate along routes suited to their reproductive and feeding needs—some of them up to 2000 km annually. Most of them spawn in late winter or early spring, each female producing many millions of eggs. The eggs are broadcast in the ocean to drift with the currents. The larvae and young, up to a few centimeters long, are also pelagic; many drift up to 3 months and hundreds of kilometers before descending to the bottom.

Like many larger fishes of cold waters, most gadoid fishes grow relatively slowly and live many years. Cod reach sexual maturity between 4 and 15 years, depending on the race, and some may live 30 years. Most of the gadoids in commercial catches are between 2 and 10 years old.

All of the gadoids feed on other animals, principally on invertebrates when small, and on fish when large. Starfishes, sea urchins, mollusks, crustaceans, and fish are common foods.

3.3.15 PERCOIDEI—PERCHLIKE SPINY-RAYED FISHES

The order Perciformes includes a large proportion of the total species of fish. These fishes usually have two dorsal fins of which the first is spiny. Their pelvic fins are usually located under or in front of the pectoral fins.

This morphologically diverse order is divided among about 20 suborders of which three are especially important for either food or angling. One of these, the Percoidei, includes about 100 families that are most like the original type of the order in that they have spines in the fins. They inhabit both marine and fresh waters ranging from tropical to subarctic, but a large proportion of the marine species are found in the coastal zone of either tropical or subtropical waters. The species of major importance are mostly those in the following eight families, which are grouped together here because they are similar in general appearance and habit. (The other especially important suborders are the Scombroidei, the tunas and mackerels, and the Cottoidei, which includes the rockfishes of the family Scorpaenidae.)

The sea basses and groupers are the best known of the more than 400 species of the family Serranidae (Fig. 3.28). They are usually robust with large mouths, small teeth, ctenoid scales, and either a round or truncate caudal fin. Most species live in shallow tropical or subtropical seas, but a few occur along temperate shores or in fresh water. Most are predatory on other fishes or invertebrates. Some may exceed 400 kg in weight. Most are

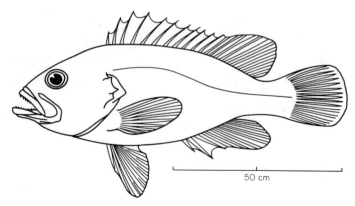

Fig. 3.28. A grouper, *Epinephelus guaza*, family Serranidae, from the Mediterranean.

excellent to eat, and many are sought by anglers. Some of the smaller tropical species can change color rapidly and are therefore called chameleons of the sea.

The snappers of the family Lutjanidae (Fig. 3.29) include about 250 species. Most are large, voracious, and brightly colored fish of shallow tropical or subtropical seas. They have a large mouth with canine teeth, robust bodies, ctenoid scales, and usually a truncate or forked caudal fin. Most are highly valued for either food or sport, and some for both. The name *snapper* has been applied because some species snap their jaws tightly together after being caught.

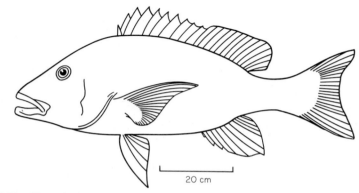

Fig. 3.29. The red snapper, or pargo, *Lutjanus aya*, family Lutjanidae, one of the excellent food fishes of the Caribbean region.

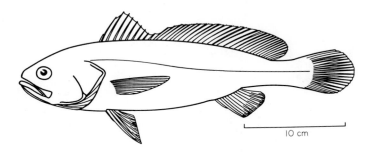

Fig. 3.30. The yellow croaker, *Pseudosciaena manchurica*, family Sciaenidae, one of the most abundant food fish of the East China Sea.

The croakers and drums of the family Sciaenidae (Fig. 3.30) comprise about 150 species, most of which live in shallow tropical or subtropical seas. They are usually sombre in color and in some tropical fisheries are the major fishes taken on the bottom in depths of 10–80 m. Many species are sought for food, a few for sport. The name *croaker* or *drum* has been applied to many of the species because they have a resonant air bladder with a vibratory mechanism that produces fairly loud croaks or grunts. These fishes usually have a medium-sized or small mouth, a second dorsal fin with a longer base than the first, and ctenoid scales. All are carnivorous.

The porgies of the family Sparidae (Fig. 3.31) include more than 100 species. These occur in seas ranging from tropical to subarctic. A few

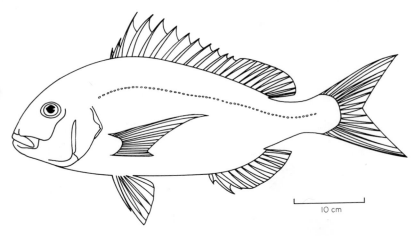

Fig. 3.31. The Japanese Ma-dai, *Chrysophrys major*, family Sparidae, one of the most delicious fish in the world.

species attain a large size; some are prized by anglers. Some species of moderate size comprise the major commercial stocks on tropical continental shelves between 60 and 200 m. Usually they have a small mouth, a compressed body, and a forked caudal fin. All are carnivorous.

The jacks, scads, and pompanos of the family Carangidae (Fig. 3.32) include about 200 species. They inhabit seas ranging from tropical to temperate. Most are good eating. Some are highly prized as game fishes, e.g., amberjacks of the genus *Seriola* and the cavalla or kingfish of the genus *Caranx*. Some are schooling fishes such as the genus *Trachurus* which includes the maasbanker of South Africa and the jack mackerel of the North Pacific Ocean and support substantial commercial fisheries. The members of the family are characterized by compressed bodies, small, smooth scales or none at all, forked caudal fins, two separate spines in front of the anal fin, and a ridge of bony plates along either side of the caudal peduncle in some species. All are carnivorous and robust, swift swimmers.

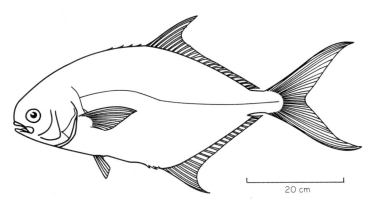

20 cm

Fig. 3.32. The pompano, *Trachinotus carolinus*, family Carangidae, one of the choicest food fish of the western tropical Atlantic.

The perches of the family Percidae (Fig. 3.33) are freshwater fishes of the Northern Hemisphere. The common perches of the genus *Perca* of North America, Europe, and Asia and the walleyed pikes of the genera *Stizostedion* in North America and *Lucioperca* of Europe and Asia are important fish for food and sport. Commonly the members of the family have a medium to small mouth, separated dorsal fins, and ctenoid scales. They are all carnivorous and feed on either small fishes or invertebrates. A great majority of the American species in the family belong to the subfamily Etheostominae, the darters. These are commonly brightly colored fish less than 10 cm long that live in running water on stony bottoms.

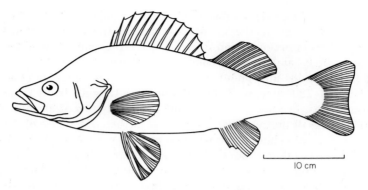

Fig. 3.33. The yellow perch, *Perca flavescens*, family Percidae, of fresh waters in Europe and North America.

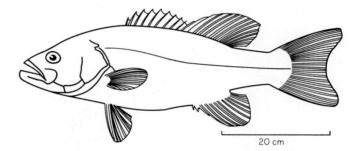

Fig. 3.34. The smallmouth black bass, *Micropterus dolomieu*, family Centrarchidae, a choice sport fish of North America.

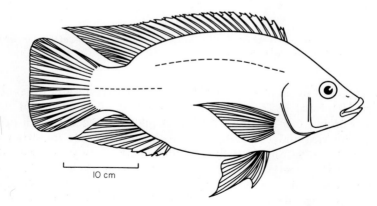

Fig. 3.35. The cichlid, *Tilapia mossambica*, family Cichlidae, from fresh waters of Africa, a species of great promise for aquaculture.

The sunfishes and freshwater basses of the family Centrarchidae (Fig. 3.34) occur in fresh water in North America. Only about 30 species are recognized but many are well known to anglers in the lakes and slow rivers from Canada to Mexico. All prefer moderate to warm waters with either sand or mud bottoms and abundant vegetation. All species use nests. The males build the nests and guard the eggs and young. All species have more or less flattened bodies, ctenoid scales, and small to medium mouths. They feed principally on either small fish or invertebrates.

Similar fishes of the family Cichlidae live in the fresh waters of Africa and South America. These include the genus *Tilapia* (Fig. 3.35), an important food fish that is of increasing use in aquaculture. The family also has many small colorful fish with unusual breeding habits (e.g., mouthbrooding) that are highly prized by aquariists.

3.3.16 SCOMBRIDAE*—MACKERELS AND TUNAS

Most scombrids are schooling marine fishes of surface waters ranging from tropical to temperate. They may migrate over the continental shelf or far outside its edge; some tunas make transoceanic migrations. All species are prolific and most broadcast their eggs in the surface layer of the open sea. Most scombrids have rich, oily flesh and are excellent eating. Some of the larger species are highly prized game fish.

The scombroid fishes are robust, torpedo-shaped, voracious fish. Some have small, inconspicuous scales; others are partly naked. The mouth is large and the color is usually blue or green on top and silvery-white underneath. Their dorsal and anal fins are followed by finlets, and their caudal peduncles are reinforced by one to several keels on the sides.

The more abundant mackerels are of the genus *Scomber* (Fig. 3.36) and are found in most temperate waters of the world. They support major fisheries on both sides of the North Atlantic and North Pacific Oceans, in the Black Sea, and in a few places in the Southern Hemisphere. They range in size up to about 60 cm, congregate in great schools, and migrate toward colder water during the spring and summer. A closely related genus that supports important fisheries is *Rastrelliger* of the Indian Ocean.

Larger (to 200 cm), swifter, more voracious, yet still slender mackerel-like fish are the Spanish, or king, mackerels of the genus *Scomberomorus*. These fish inhabit tropical and subtropical waters. They are prized as choice

---

*Taxonomists agree that the mackerels and the tunas are closely related and commonly refer to them as the "scombroid fishes." They disagree however as to whether the groups should be subfamilies, families, or even suborders. Here they are considered as subfamilies of the family Scombridae.

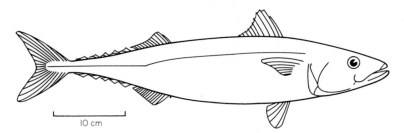

Fig. 3.36.   The Atlantic mackerel, *Scomber scombrus*, family Scombridae.

food and game fish but seldom are taken in large quantities because they tend to migrate either as individuals or in small schools.

Three genera are intermediate in characteristics between the mackerels and tunas: *Auxis*, the frigate mackerels; *Euthynnus*, the little tunas; and *Sarda*, the bonitos. These fish range in size from small to medium (to about 110 cm). *Euthynnus* (*Katsuwonus*) *pelamis*, the skipjack tuna, is an extremely valuable species found around the world in tropical and subtropical waters, but the other species in these genera are only locally important.

The large scombrids, the big tunas of the genus *Thunnus*, are the objects of major fisheries around the world in waters ranging from tropical to temperate. The largest species, the bluefin, *T. thynnus*, reaches a length of 3.5 m and a weight of more than 800 kg. It tends to be more abundant near the coasts on both sides of the Atlantic and Pacific in subtropical to temperate waters than elsewhere in its distribution. Recoveries of tagged individuals have shown transatlantic crossings. The other species of temperate waters is the albacore, *T. alalunga*. It seldom exceeds 120 cm in length but is especially valuable for canning because of its white flesh. This species has

Fig. 3.37.   The yellowfin tuna, *Thunnus albacares*, family Scombridae, of the tropical and subtropical oceans.

been shown to make transpacific migrations. The species that prevail and support large fisheries on the tropical and subtropical high seas around the world are the yellowfin tuna, *T. albacares* (Fig. 3.37), and the bigeye tuna, *T. obesus*. These species reach a length of about 180 cm.

3.3.17 SCORPAENIDAE—SCORPION AND ROCKFISHES

Many of the more than 250 species of fish in this family are considered to be among the ugliest of the fishes. They are distinguished by large, heavy, bony heads with spines and a heavy bone extending along the cheek from below the eye. They have small teeth and ctenoid scales. Most species range in size from small to medium, rarely exceeding 100 cm in length.

All of the scorpaenids are marine fish occurring in waters ranging from shallow to deep and from the arctic to the tropics. Most species prefer rocky areas, hence the name *rockfishes*. Many of the species are brilliantly colored and many have fleshy flaps on their fins which effectively camouflage them among rocks or reefs. All are carnivorous. Many species, including those of greatest commercial importance, are ovoviviparous. The eggs are fertilized internally and the eggs and larvae are contained within the females until the larvae are spawned.

The family is commercially important for two reasons. Many species have venom sacs at the base of their fin spines and can inflict painful wounds when handled. Two genera of the tropical Pacific, *Synanceja** and *Pterois*, produce wounds that may be fatal to humans.

Despite the ugly appearance and venomous equipment of some, almost all of the larger species are choice eating. Of most value are the three species of *Sebastes* (Fig. 3.38), the ocean perch or rosefish of the North Atlantic, and the more than 60 species of *Sebastodes*, the rockfish of the North Pacific. The species of both of these genera are the objective of extensive trawl and long-line fisheries.

3.3.18 PLEURONECTIFORMES—FLATFISHES

The more than 500 species of the order Pleuronectiformes are asymmetrical fish when adult. Their bodies are flattened from side to side. They swim upright when in the larval stage and for a time thereafter, but usually swim on one side later. As they begin to swim on one side an eye migrates to the "top" side and the top side becomes colored while usually the "bottom" side remains white. They are anatomically similar to the perchlike fishes but commonly lack spiny rays in their fins.

Synanceja is placed in another family by some recent authors.

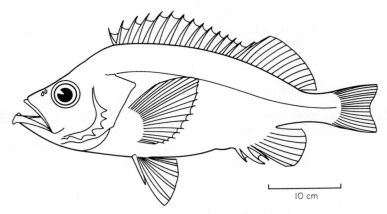

Fig. 3.38. The rosefish, *Sebastes marinus*, family Scorpaenidae, a major food fish of the North Atlantic.

The order is usually divided into seven families of which only three are of commercial importance: Pleuronectidae (Fig. 3.39) the righteye flounders; Bothidae (Fig. 3.40), the lefteye flounders; and Soleidae, the soles. The English common names are not consistent with the family affiliations, however. The name *sole* is applied to the thinner species of all three families. The names *dab*, *flounder*, *plaice* are used for the medium-sized species, and the name *turbot* is applied to the larger robust species of the first two families listed. The halibuts, the largest of the flatfishes (reaching more than 200 kg and 3 m), are in the family Pleuronectidae.

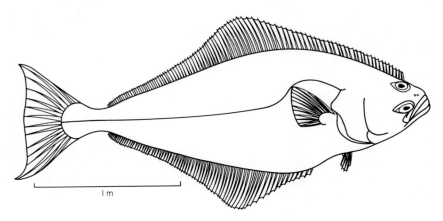

Fig. 3.39. The halibut, *Hippoglossus hippoglossus*, family Pleuronectidae, of the North Atlantic.

The flatfishes live in closer association with the bottom than most other fish. They occur from the tropics to the arctic, but the more important species are in temperate and arctic waters of the Northern Hemisphere. They are all marine but a few species migrate up rivers into fresh water. They live either on the continental shelf or on the upper part of the continental slope. They feed mostly on either invertebrate animals or small fish. Commonly they rest on the bottom changing their color and color patterns to match it. They spawn on the bottom, but the eggs and larvae are usually pelagic, drifting with the ocean currents. The postlarvae are pelagic only until they turn on their sides when they descend to the bottom.

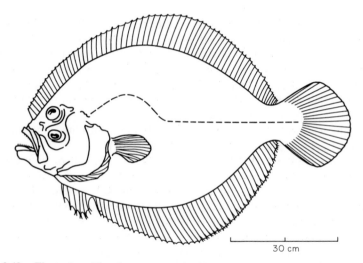

30 cm

Fig. 3.40. The turbot, *Rhombus maximus,* family Bothidae, of the Northeast Atlantic.

Many species are highly esteemed for their nearly boneless, delicately flavored flesh. They are caught usually by trawls and are associated frequently with codlike fish.

3.3.19 CETACEA—WHALES

All of the larger marine mammals have been sought for either food, oil, or pelts by coastal peoples in many parts of the world, but in recent years the large cetaceans, the whales, have been the species of most value. These include the largest of the toothed whales, the sperm whale, and several species of rorquals, one of the groups of baleen whales. The right whales, another group of baleen whales, and many of the numerous species of porpoises and dolphins are taken in minor quantities.

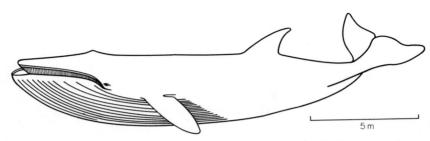

Fig. 3.41. The sei whale, *Balaenoptera borealis*, family Balaenopteridae, now the most abundant of the baleen whales.

The rorquals are the most slender and fastest of the baleen whales. They have short baleen plates and a small dorsal fin. They include the famous blue whale (the largest living animal which surpasses a length of 30 m and weight of 120 tons), the fin whale, the sei whale (Fig. 3.41), and the humpback whale. These mammals have a lifespan of about 30 years during which time each female produces about 10 large young each measuring about one-fourth the length of a mature animal.

The right whales are large, slow baleen whales with large jaws and long baleen plates. They are no longer abundant.

The baleen whales bear their young either in tropical or in subtropical waters. Food is usually scarce in these waters; hence most baleen whales migrate towards the poles in the spring to feed and fatten during the summer. The small crustaceans such as krill or small fish on which they feed are abundant in summer in the far northern or far southern oceans. They return to breed, have their calves, and commonly fast during the winter in warmer waters. On the other hand, the sperm whale (Fig. 3.42) feeds largely on squid and it finds food abundant in some tropical waters as well as in the higher latitudes. Unlike the baleen whales it is more widely distributed in tropical, subtropical, and temperate latitudes; only a few solitary males go to the edges of the ice—both north and south.

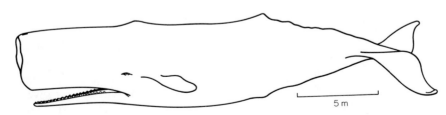

Fig. 3.42. The sperm whale, *Physeter catadon*, family Physeteridae, the largest of the toothed whales.

The whale fisheries of the world developed near land along which the whales had regular migratory routes and where men in small boats were able to catch some of the slower species, the *right whales*. Whaling spread farther at sea in the fifteenth to seventeenth centuries as men took to sailing ships to seek them for oil and whalebone. Arctic whaling for the right whales became a major pursuit in the seventeenth and eighteenth centuries and resulted in a depletion of the stock by the middle of the eighteenth century. Tropical whaling, especially for the sperm whale, developed in the eighteenth and nineteenth centuries but declined after 1850 for economic reasons (not a scarcity of sperm whales).

Throughout these centuries most of the fast rorquals escaped the whalers, but, with the development of fast catcher ships and harpoon guns in the early twentieth century, these species became obtainable. The fishery expanded and severely reduced first the Arctic and then the Antarctic stocks. The blue whale is now nearly extinct; the fin whale is very scarce. The present whale fishery of the world produces mainly sei whales and sperm whales.

REFERENCES

The literature that describes and classifies animals and plants is immense. Descriptions, keys, and check lists of major groups have been prepared for most localities of the world and should be sought by the scientist. Only a few general or comprehensive regional reference works are given here.

Bailey, R. M., chm. (1970). "A List of Common and Scientific Names of Fishes from the United States and Canada," 3rd ed., Spec. Publ. No. 6. Amer. Fish. Soc., Washington, D.C.

Berg, L. S. (1940). Classification of fishes, both recent and fossil. (In Russian.) *Trav. Inst. Zool. Acad. Sci. URSS* **5**(2), 87–345 (reprint with English transl., Edwards, Ann Arbor, Michigan, 1947).

Bigelow, H. B., and Schroeder, W. C. (1953). Fishes of the Gulf of Maine. *U.S., Fish Wildl. Serv., Fish. Bull.* **53**(74), 1–577.

Blackwelder, R. E. (1967). "Taxonomy." Wiley, New York.

Clemens, W. A., and Wilby, G. V. (1961). "Fishes of the Pacific Coast of Canada," (2nd ed.) *Bull. Fish. Res. Bb. Can.* **68**, 443.

Dawson, E. Y. (1966). "Marine Botany." Holt, New York.

Edmondson, W. T., ed. (1959). "Fresh-Water Biology," 2nd ed. Wiley, New York.

Greenwood, P. H., Rosen, D. E., Weitzman, S. H., and Myers, G. S. (1966). Phyletic studies of teleostean fishes with a provisional classification of living forms. *Bull. Amer. Mus. Natur. Hist.* **131**(4), 339–456.

Hardy, Sir A. (1965). "The Open Sea: Its Natural History," Part 1. The World of Plankton; Part 2. Fish and Fisheries. Houghton, Boston, Massachusetts.

Jordan, D. S., and Everman, B. W. (1896–1900). The fishes of North and Middle America. *Bull. U.S. Nat. Mus.* **47**, 1–3313.

Lagler, K. F., Bardach, J. E., and Miller, R. R. (1962). "Ichthyology, the Study of Fishes." Wiley, New York.

Lowe-McConnell, R. H. (1968). Identification. *In* "Methods for Assessment of Fish Production in Fresh Waters (W. E. Ricker, ed.), Int. Biol. Progr. Handb. 3, p. 46–77. Blackwell, Oxford.

Matsubara, K. (1955) "Fish Morphology and Hierarchy," Vol. 1. Ishizaki-Shoten, Tokyo. (In Japanese.)

Matsubara, K. (1967). "Fish Morphology and Hierarchy," 2nd ed., Vols. 2 and 3. Ishizaki-Shoten, Tokyo. (In Japanese.)

Mayr, E. (1963). "Animal Species and Evolution." Harvard Univ. Press, Cambridge, Massachusetts.

Nikol'skii, G. V. (1961). "Special Ichthyology (Chastnaya ikhiologiya)" (transl. from Russian), 2nd rev. enlarged ed. Israel Program Sci. Transl., Jerusalem.

Schmitt, W. L. (1921). The marine decapod crustacea of California. *Univ. Calif., Publ. Zool. Berkeley* **23**, 1–470.

Sears Foundation for Marine Research. (1948–1971). Various authors. "Fishes of the Western North Atlantic." Sears Found. Mar. Res, New Haven, Connecticut.

Slijper, E. J. (1962). "Whales." Hutchinson, London. (A translation of "Walvissen" published in the Dutch language in 1958 by Centen, Amsterdam.)

Smith, J. L. B. (1950). "The Sea Fishes of Southern Africa." Central News Agency Ltd., South Africa.

Sterba, G. (1962). "Freshwater Fishes of the World." Longacre Press Ltd., London. (A translation by Denys W. Tucker of "Susswasserfische aus aller Welt," Zimmer, Berchtesgaden, 1959.)

Stoll, N. R., chm. (1964). "International Code of Zoological Nomenclature," adopted by 15th Int. Congr. Zool. Int. Trust for Zool. Nomenclature. London.

Svetovidov, A. N. (1962). "Gadiformes (Treskoobraznye)" (transl. from Russian by W. J. Walters with V. Z. Walters). Israel Program Sci. Trans., Jerusalem.

Trautman, M. B. (1957). "The Fishes of Ohio." Ohio State Univ. Press, Columbus, Ohio.

Yonge, C. M. (1949). "The Sea Shore." Collins, London.

Biology of Aquatic Resource Organisms

Increasingly, fishery scientists are being asked questions such as the following: How does a fish find a lure? What temperature, salinity, and oxygen concentration limit the distribution of an acquatic animal? How does an animal navigate on the high seas? What is the best place and time to catch tuna? How and why do fish form schools?

Answering such questions requires first an understanding of how the individual organism lives in the aquatic environment, how it senses its surroundings, how it breathes, eats, excretes, reproduces, and develops. Every organism performs these functions in ways to which it is suited by structure and habit; therefore, the biologist studies separately the structure, physiology, and ethology (behavior). In this chapter the principal functions of certain aquatic organisms are summarized, with emphasis on the behavioral aspects that are useful in answering questions such as those stated above.

4.1 Senses

Every skilled angler knows how easily a trout is alarmed by seeing or hearing him. Every purse seine fisherman knows the frustration of having a school of fish take alarm and escape his net before he can close it. Each approaches his quarry knowing that his presence can be sensed through vision, hearing, or in some other way. When close enough, the fisherman may try to lure the fish to bite or enter a net. He may employ either a visual stimulus such as an artificial lure or light, or a chemical stimulus such as a natural bait or sexual product. He may frighten fish into his net. In most of his strategy the fisherman depends on overcoming the quarry's senses and behavior.

4.1.1 LIGHT RECEPTION

Many of the simpler invertebrate animals can sense light and will orient themselves to light from one direction. The light-sensitive organs range in complexity from light-sensitive spots to camera eyes with lenses capable of focusing an image on a retina. Crustacea have compound eyes with many tubelike units, each of which receives light from a different direction; these animals probably receive a mosaiclike image. Many bivalve mollusks have simple eyes with lenses; some have more complex eyes.

Camera eyes occur in the cephalopod mollusks and in most fish. Most such eyes have mechanisms for focusing on either near or distant objects and for controlling the amount of light. The latter may be either an iris near the lens or movable cells in the retina that put pigment in front of the visual cells.

Even with the best of eyes, underwater animals are severely restricted in vision by the particles in the water. The limits are about the same as those of the Secchi disc transparency which are about 60 m in the clearest of ocean waters and only a few centimeters in turbid coastal waters. Some of the best eyes among fish are probably possessed by the tunas; they live in the clearer ocean waters and communicate among themselves with visual signals. Tuna living in experimental tanks were able to see brightly illuminated black and white stripes that subtended an angle as little as 4 minutes of arc.* Under the same excellent conditions a net twine. 1 mm in diameter could not be seen at a distance of more than 1 m.

Vision may be better when the eye of the animal can perceive colors but the poor transparency of water restricts color vision at the red end of the spectrum. Water best transmits the blue–green part of the spectrum (400–520 mμ). Fish eyes appear to be most sensitive to these colors and the bioluminescent organs produce the brightest light within this range.

The image of an object in the air when seen from below the water surface is restricted severely by the refraction of light at the surface. When the surface is disturbed, as it almost always is, the image is broken into many moving parts. Even with a calm surface, the image of everything above the water from horizon to horizon is restricted to a cone, or Snell circle, subtended by an angle of 97.6° (Fig. 4.1). Objects near the horizon are visible but greatly distorted.

---

*E. L. Nakamura (1968). Visual acuity of two tunas, *Katsuwonus pelamis* and *Euthynnus affinis*. *Copeia* No. 1, 41–49.

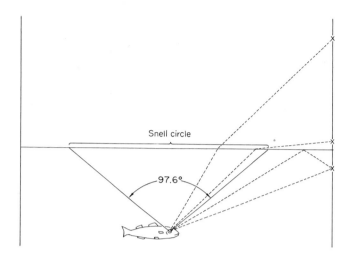

Fig. 4.1. Diagram of the field of vision from below the water surface.

Some fish are attracted by light, and fishing is conducted at night with lights in many parts of the world. Usually a powerful lamp is hung over the side of a stationary ship, either below or above the surface, until the fish are congregated in such a way that they may be encircled by a net. Several species of herrings, anchovies, mackerels, tunas, and sauries (all schooling fishes) are taken by this method. Perhaps most spectacular is the technique used in the fishery for sprat in the Caspian Sea; the fish are attracted by lights to the mouth of a suction hose and merely pumped aboard. It is not known whether fish are attracted directly by light or by their ability to see food in the lighted area, but in either case they are responding to a visual stimulus.

Animals that either school (see Section 4.4) or orient to each other probably rely principally on vision; some of these animals communicate with visual signals, such as the tunas. Some animals develop sexual differences in color and can change color rapidly during mating, perhaps as a way of communicating.

Migratory aquatic animals (like numerous terrestrial animals) may orient to the sun. Evidence of this behavior was gathered from observations in the field and from experiments in training fish to respond to a visual cue for a reward. Fish that had been moved from their home and marked with small floats or sonic tags were found to follow a consistent compass direction when the sun was shining and to become confused when the sun was obscured. Trained captive fish were observed to follow a definite compass course using the sun for guidance and compensating for the change in the azimuth of the sun as it passed through the sky. Some further evidence indicates that fish can also change their orientation according to the maximum altitude of the sun (Hasler, 1966). Such orientation means not only a remarkable ability to "measure" the angle of the altitude of the sun from below the surface of the water, but also a fairly accurate sense of time, both daily and seasonal.

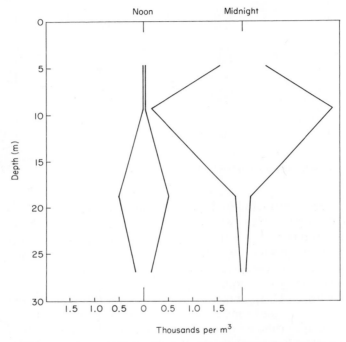

Fig. 4.2.   Vertical distribution of calanoid copepods in Lake Nerka, Alaska, during a summer day. Data from D. E. Rogers (1968). A comparison of the food of sockeye salmon fry and three-spine sticklebacks in the Wood River lakes. *Univ. Wash. Publ. Fish.* [N.S.] **3**, 1–43.

Another response to light is evidenced by the extensive daily vertical migrations of many plankton animals. This behavior was discovered by some of the first people to tow plankton nets when they caught much more plankton near the surface at night than during the day (Fig. 4.2). More observations on the prevalence of the phenomenon were collected by the operators of sensitive echo-sounding equipment; they obtained records of strong echos from numerous layers that rose toward the surface in the evening and descended in the morning. They called them "deep scattering layers" (Fig. 4.3). At first the cause of the echos was a mystery, but later it was found to be layers of plankton of many kinds, some of which were attracting fish.

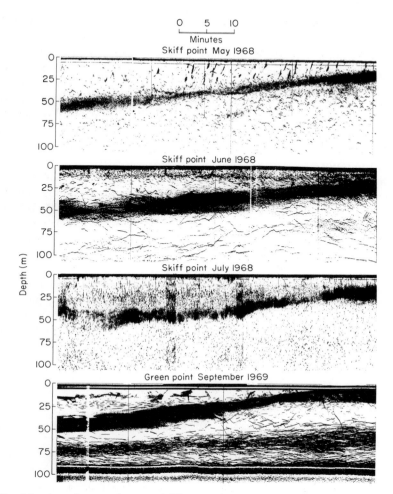

Fig. 4.3. An echo sounder on a drifting vessel in Puget Sound repeatedly recorded the vertical migration of euphausids during the hour before sunset from depths of about 50 m up to a depth of about 25 m. (Records by R. T. Cooney, Dept. of Oceanography, University of Washington, Seattle, Washington.)

4.1.2 CHEMICAL RECEPTION

In terrestrial animals, taste receptors come in direct contact with a substance and smell receptors detect substances at a distance; both senses require that the substance be dissolved in water of the mucous membrane. In aquatic animals the substances are dissolved in water at all times, but it is useful to retain the terms *taste* for contact reception and *smell* for distance reception even though the organs may not be in the mucous membrane in the mouth or nasal area.

Taste and smell may be the primary senses of many aquatic animals for locating food, home, mates, and enemies even if the animals have good eyes. In some fish apparently the ability to smell is at least equal to that of the most sensitive of terrestrial animals. It is put to good use in waters where the visibility is only a few centimeters. Usually taste and smell organs are associated with the feeding mechanisms, but in many animals, including fish, they are associated with sensitive chemical receptors located in other parts of the body.

Such senses are probably especially useful to aquatic animals because most waters have odors resulting from animals and plants, some of which are detectable even by relatively insensitive human noses. Many algae cause odors that are troublesome or even offensive when the water is used for either drinking or swimming. Many live fish have characteristic odors and some coastal fishermen can detect schools of fish in the water by the odor.

The olfactory organs of most fish are cavities lined with sensitive cells through which water is circulated. Some of their olfactory organs and the associated parts of the brain are so large that the fish would be expected to have an excellent sense of smell. An example is the haddock (Fig. 4.4).

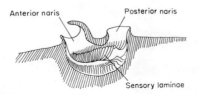

Anterior naris   Posterior naris

Sensory laminae

Fig. 4.4. A longitudinal section of the olfactory organ of the haddock *Melanogrammus aeglefinus*, [Redrawn from R. H. Burne (1909). The anatomy of the olfactory organ of teleostean fishes. *Proc. Zool. Soc. London* pp. 610–663.]

Salmon use the sense of smell for homing. This was established by moving adult salmon from their spawning streams to locations some miles away and surgically interfering with various senses. Those with their eyes covered quickly found their way back to the stream from which they had been taken, but those with plugged olfactory organs became lost. In other experiments salmon with normal olfactory organs found their own home stream even though they had been mixed with salmon in another spawning stream, whereas those with their olfactory organs plugged did not return. Clearly, the home stream had a distinctive odor that they could recognize.

Salmon can detect the odor of enemies. A spectacular escape reaction will be caused in a group of resting adult salmon in a stream if a bear wades some distance upstream. The salmon will swim wildly downstream when they come in contact with the water in which the bear has walked. The same reaction would be obtained if a person were to rinse his hands in the water upstream from the salmon.

Finding and sorting food is probably the predominant role of the taste and smell senses. Many experiments have shown that aquatic animals react to extracts of food. They sense the food by their olfactory organs; some also sense the food through special receptors on fins and barbels. The ability of sharks to detect blood in the water and their subsequent ferocity in a feeding frenzy is notorious.

Chemical receptors play a major role in the spawning of shelled mollusks. The oyster is an example. The male oyster may be stimulated to spawn either by rising temperature, by the sperm of other oysters of the same species, or by a wide variety of organic substances. The females are more specific; they spawn only when stimulated by the sperm from the males. An entire oyster population will spawn soon after the first male spawns because his sperm will start a chain reaction with other males and they in turn will stimulate the females.

The possibilities of taking advantage of the chemical sensitivity of fish and other aquatic animals are only beginning to be scientifically studied. There will be many applications for effective attractants and repellants that can be used in catching, guiding, and protecting aquatic animals.

4.1.3 TEMPERATURE RECEPTION

Most mobile aquatic animals seem to prefer certain temperatures; each species is commonly found only in waters with temperatures within a certain range; many migrate seasonally according to the warming and cooling of the water. One may assume that they can detect temperature differences and respond accordingly.

However, the role and method of temperature reception are not well understood. Fish respond to temperature changes as small as 0.03°–0.07°C when the changes occur fairly rapidly, but how such responses can guide migrations is not clear. Neither is it clear how temperature changes are perceived; some evidence indicates that taste organs and the lateral line system are involved.

4.1.4 MECHANICAL RECEPTION

All of the higher aquatic animals respond to mechanical stimulation; some are extremely sensitive to touch. The use of touch in sensing their surroundings is universal among animals and needs no further emphasis, but

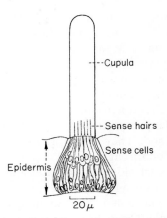

Fig. 4.5. A sensory unit of the lateral line system. These units are located either on the surface of the body or in lateral line canals. [After S. Dijkgraaf (1962). The functioning and significance of the lateral line organs. *Biol. Rev.* **38**, 51–105.]

one related function is only common to most fish. This is the ability to sense water currents and low-frequency vibrations (less than about 200 Hz) through the lateral line system. This is a system of furrows or canals with frequent openings to the exterior that occurs along the trunk and on the head. It has specialized sensory cells at frequent intervals (Fig. 4.5).

The lateral line appears to be a set of distant touch organs that allow the fish to sense the currents set up by movements of nearby animals and the presence of nearby objects by reflected currents from his own movements. The distribution of sensory cells along the line permits identification of the direction of a point source of stimuli. The lateral line apparently supplements the sense of hearing at low frequencies. In fact, there is no good dividing line between high- and low-frequency reception and there is some confusion about near-field and far-field effects.

4.1.5 PRESSURE RECEPTION

All swimming or drifting aquatic animals must regulate their depth and must be expected to respond to pressure changes. Animals having an air sac, such as fish with an air bladder, have an organ that will change in size according to pressure. Surprisingly, many invertebrates that lack such organs are also pressure sensitive. Some crustaceans as well as some fish have been shown to respond to pressure changes of only 10 mbars, the equivalent of a 10-cm change in the depth of water. It has been postulated that pressure receptors may be developed either around changes in electric potential with pressure across the respiratory surfaces of crustaceans or with differences in compressibility of skeletal and soft tissues. Pressure receptors may be combined with mechanical receptor cells such as those in the lateral line canals of fish.

4.1.6 SOUND RECEPTION

One might expect the aquatic animals to depend more on hearing for information than the terrestrial animals because sound travels much faster and with less loss of energy in water than in air. Such does not seem to be the case; however the role of sound reception is not well understood and involves some activities that are rare among terrestrial forms. Although a majority of the fish and many crustaceans produce sounds of one kind or another (see Section 2.3.4) only a few fish and the marine mammals appear to produce sounds for communication.

The simplest hearing organs in some crustaceans are mechanical receptors; small, projecting, hairlike organs that are sensitive to deflection. Similar structures occur in the lateral line canals of fishes. Such skin receptors are sensitive only to low-frequency vibrations (up to about 200 Hz) or to water currents. An inner ear, or labyrinth, that is similar to the mammalian ear enables fish to detect frequencies higher than about 50 Hz.

The ability of fish to hear sounds of different frequencies has been studied in behavioral experiments in which fish have been rewarded by responses to various frequencies. These studies have shown that the most sensitive frequency range for most fish is 300–800 Hz with about 2000 Hz as the upper limit. Members of the Cypriniformes, a group of freshwater fish with a special connection between the ear and air bladder, can detect frequencies up to about 13,000 Hz (the upper limit for humans is about 10,000 Hz).

When fish do communicate by sounds they are usually either feeding or breeding. The sounds may be helpful in locating mates, in protecting nests against predators, or in telling others about food. The few attempts to attract fish by sounds were successful only when the natural sounds of food or breeding were imitated. Some fish react strongly to unusual sounds with an escape reaction of leaping or diving away from the sound. A few fish appear to produce sounds for the purpose of echo ranging.

Vastly more sophisticated is the auditory capability of the marine mammals. Some of these produce and hear sounds up to 150,000 or even 170,000 Hz. They have a substantial vocabulary for communication among themselves and make extensive use of echo ranging. Furthermore they can locate the direction of a sound as their two ears are acoustically isolated.

Despite the ability of most fish to hear, and despite the alarm reactions of some fish, the use of sound to repel fish has been notably difficult. Experiments with frequencies from 67 to 70,000 Hz and intensities high enough to be described as underwater thunder failed to cause young trout to do more than to give an initial start at the commencement of the sound.* Other workers who studied the effects of sound on fish entering trawl nets were unable to show that the sound of nets and towing wires has a significant effect either by helping the fish avoid the nets or by attracting them to the nets. Still others were able to make sounds that tended to attract the fish toward the bottom where they could be caught more easily by trawls. Perhaps the sound would have to be one that the fish associates with danger.

4.1.7 ELECTRICAL RECEPTION

Most of the studies about the effect of electricity on fish have been either attempts to understand the use of electric organs by the species of fish that produce electricity or attempts to guide and catch fish by electricity. Little has been done to identify how electricity is received, although it was shown that some special electrical receptors appear to occur in the lateral line of the electric eel, *Electrophorus*. Certain sensory and motor nerves in fish are direct electrical receptors, and presumably, the entire body will receive electricity.

The production of electricity occurs in a few species of both freshwater and saltwater fish of widely diverse families. These include some marine torpedos and rays (Elasmobranchs), and some freshwater Mormyridae of Africa, Electrophoridae and Gymnotidae of South America—the latter families being related to the minnow and the carp family, Cyprinidae. The electric organs are modified muscles. The most powerful of these fishes appears to be the marine torpedo; its maximum discharge is between 3 and 6 kW. The torpedos and rays use their powerful electric organs to stun prey, whereas the freshwater fishes use their much weaker electric organs for orientation as well as to stun prey. Some of the latter have a remarkable ability to sense differences in the resistance of the water around them. Small wires or nonconductive pieces of glass tubing when moved near them bring quick responses. They can detect differences in electric potential of as little as 0.02 μV/cm.

Salmon have been found to move predominantly downstream with the ocean currents in their oceanic migrations. Since the electric potential in such ocean currents (ranging from 0.05 to 0.5 μV/cm, see Section 2.3.5) is more than that detectable by some electric fish, it has been postulated that salmon might utilize the electric field of the ocean for orientation during their long migrations.

---

* C. J. Burner and H. L. Moore (1962). "Attempts to guide small fish with underwater sound." *U.S., Fish Wildl. Serv., Spec. Sci. Rep.—Fish.* **403**, 1–30.

When electric currents through the water are greatly and continuously increased, fish display a characteristic series of reactions. When the current is alternating in polarity, at first the fish is uneasy, then it gives a "minimum response" (a powerful twitch of its tail), then twitches violently, and then becomes paralyzed. When the current is direct, the fish becomes uneasy, then faces the positive pole and swims toward it (electrotaxis), then becomes paralyzed (electronarcosis). When the direct current is interrupted at a suitable frequency (but not alternated in polarity), the fish will swim vigorously toward the positive pole during electrotaxis.

The levels of electric potential at which fish behave in the above manner vary according to temperature, electrical resistance of the water, kind of electric current, size and species of fish, but the general levels are somewhere between 0.01 and 1.0 V/cm. The responses occur at lower levels of electric potential with lower temperatures, in seawater, with alternating current, and with larger fish.

Such behavior of fish in tanks has suggested to many workers the exciting possibility of forcing fish by electricity to swim into nets or away from water intakes. Unfortunately, such hopes have been dashed by the difficulty of maintaining a uniform electric field outside the laboratory and by the huge power required in seawater. As fish enter any electric field some may be repelled before being compelled to swim toward the positive pole; others may be narcotized before reaching the positive pole. Further, the effect of the electric field varies with the orientation of fish relative to the field and the resistance of nearby things such as the ship hull or the bottom. Nevertheless, considerable success has been achieved by the use of interrupted direct current and by the combination of electricity with other methods of concentrating fish, such as lights or nets.

4.2 Respiration

The respiration of animals includes the exchange of gases either at the surface of the body or in internal passages, the transportation of gases through the body, and the exchange of gases between the circulatory system and the cells. The gases are the needed oxygen, the waste carbon dioxide, and the inert gases, such as nitrogen, that just ride along.

The exchange of gases at the surface of aquatic animals takes place through the integument, the gut, gills, lungs, and air sacs. In tiny animals and the eggs and larvae of larger animals, the integument itself is adequate, but in larger animals larger surfaces are needed. In the more efficient systems these surfaces are supplied abundantly with blood in order to allow the rapid diffusion of gas to and from the external medium. Most fish, mollusks, and crustaceans have gills in internal chambers through which water is pumped to increase still further the efficiency of the exchange. Some have auxiliary systems for exchanging gases directly with the air either when the water is low in oxygen or when the water recedes during either tidal or annual cycles. A surprising number of fishes that live in tropical fresh waters can utilize air when living in water almost lacking in oxygen.

Oxygen consumption varies enormously with species, size, activity, season, and temperature. Fast-swimming animals use much more oxygen during active periods than sluggish animals use during periods of rest. Large animals use less per unit of weight than small animals do. Consumption increases with temperature up to some critical level then falls off, but many animals compensate for seasonal changes in metabolism and may actually consume more oxygen in winter than during the rest of the year.

The fish that require the greatest amount of oxygen are the active species that live habitually in well-oxygenated water. An example of a marine species is the mackerel; it has a relatively large gill surface and its blood will carry more oxygen than that of most fish; it needs to swim almost continuously in order to ventilate adequately its gills even though the water in which it lives is always almost completely saturated with oxygen. Freshwater species with large oxygen requirements include the trouts. These fish live habitually in the oxygen-saturated water of cold streams; they cannot long withstand water with less than 5 mg/liter of oxygen. At the other extreme (excluding the air-breathing fishes) are the less active bottom species of freshwater fish, such as carp and catfish. These fish can survive with dissolved oxygen of only 0.5 mg/liter.

Expression of such minimum levels of oxygen is an oversimplification, however, because of the synergistic action of many factors. Increasing temperatures require an increase in the amount of oxygen required to saturate the blood. Increasing carbon dioxide reduces the ability of hemoglobin in the blood to carry oxygen and therefore requires higher environmental dissolved oxygen for equivalent survival. Other related factors are the activity of the fish, its general health, and its acclimitization to the unusual temperature and gas conditions. The last is especially important because many fish have a substantial ability to adjust to conditions that are less than favorable.

Respiration in the shelled mollusks occurs simultaneously with feeding (see Section 4.5). As water is pumped through the open shell and over the gills, gas exchange occurs and food particles are filtered out. Usually the water circulated is greatly in excess of that needed for respiration alone so the amount of oxygen removed is relatively small.

4.3 Osmoregulation

The membranes of aquatic animals that pass gases and digested food between the environment and the body also pass water and salts. Most aquatic animals maintain in their bodies a reasonably constant proportion of water and salts. When it is different from the surrounding water, it requires some mechanism for regulation.

Most invertebrates in the sea have nearly the same salt concentrations (are isosmotic) as the environment. Sharks and rays tend to have a slightly higher concentration of salts (including urea) than seawater, but the fluids of most teleosts (bony fish) have much lower salt concentrations than seawater. In fresh water the fluids of all fish have a much higher salt concentration than the water. Freezing point depressions of several fluids are as tabulated below:

Seawater	-1.8 to $-2.1°C$
Shark blood	-1.9 to $-2.4°C$
Marine teleost blood	-0.7 to $-1.0°C$
Freshwater teleost blood	-0.45 to $-0.60°C$
Fresh water	-0.0 to $-0.1°C$

Marine teleosts, which lose water osmotically across the gills and must get rid of salt, tend to drink large quantities of seawater and produce only small quantities of urine. Freshwater fish, which gain water osmotically and must conserve salt, drink little or no water and secrete large quantities of very dilute urine. The salt–water balance is maintained by the kidneys, the gills, and intestine.

The different salts from the water do not remain in the body in proportion to their abundance in the environment. Most animals are able to concentrate some ions while rejecting or excreting others. Marine animals usually reject magnesium and sulfate while freshwater animals usually concentrate sodium, potassium, and chloride ions.

The mechanisms for salt control and ionic adjustment of most animals are so inflexible that the animals are stenohaline, that is, they are restricted either to salt water or fresh water. Marine fish usually tolerate salinities less than the salinity of the open sea but higher than the salinity of their blood, whereas freshwater fish usually tolerate any salinity less than their blood salinity. Relatively few estuarine animals and migratory fish, such as salmon, can adjust to the change from fresh water to salt water, or vice versa.

4.4 Schooling

It is no accident that most of the world fisheries depend on concentrations of fish rather than on fish scattered through the water. Fortunately for those who seek them, many mobile aquatic animals aggregate or school in ways that make them much easier to catch in certain places or at certain times. They school in order to breed, not only to bring the sexes together but also to release the eggs or larvae in a favorable place. They commonly school in order to feed, either because the food is localized or in order to feed collectively.

They also school because of other kinds of mutual attraction that are largely independent of environmental circumstances. Biosocial aggregations vary from casual, temporary, and irregular groupings of a few animals to very large numbers of a single species of nearly the same size, spaced uniformly, swimming parallel, and acting concomitantly, a behavior that is called "polarized" (Fig. 4.6). The variety of kinds of aggregations and the variety of causes have led to substantial disagreement on the meaning of terms and to unwieldy classification of types of schools. Some of the senior behavioral scientists who are studying fish schools now restrict the term

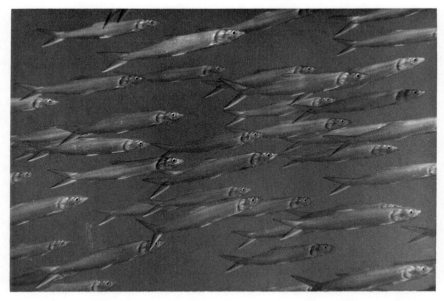

Fig. 4.6. A polarized school of machete, *Elops affinis*, in the southern Gulf of California. (Photo by Edmund S. Hobson, Bureau of Sport Fisheries and Wildlife, U.S. Dept. of the Interior. Washington, D.C.)

"school" to groups that exhibit polarized behavior. Other kinds of groups are called "aggregations," "shoals," or "clusters."

The survival value of schooling probably varies among the animals concerned. There are obvious values in finding mates and in herding food. Some fish have been shown to experience less stress when living together than when living alone. A school may give the individual a greater chance to survive because the school is harder for a predator to find, and having found it, a predator can consume only its capacity before it becomes satiated and loses the school. A school may intimidate smaller predators, whereas a solitary individual may not.

Regardless of the biological benefits of aggregating or schooling, many commercially important species do it. The herrings, anchovies, mackerels, and tunas school (in the scientific sense) near the surface of the water; the cods and many perchlike fishes aggregate near the bottom.

4.5 Food and Feeding

Every animal requires energy for living—for growth, maintenance, and reproduction—which it must obtain from its food. Each starts life with a bit of food received from a parent, but it soon needs to feed itself. It must continue to feed itself regularly with suitable food or die. The regularity must suit the animal's ability to find and ingest food and to store energy. For some, feeding is almost continuous, for others, feeding is in circadian (i.e., daily or tidal) or seasonal cycles; but for most animals feeding is the dominant activity of their entire lives because their need is constant and the food is usually scarce.

Understanding this activity is useful to any scientist concerned with any aspect of the fisheries. If he should want to improve the catch, he either would need to develop better baits or learn about feeding behavior. Should he want to develop a rational method of exploiting a population, he would need to know how food is a limiting factor and how it may be divided among competing animals. Should he culture animals, he would need to study intensively the nutritional requirements in order to obtain the best growth at the least cost.

4.5.1 FEEDING MECHANISM AND PREY

Not only does every animal feed, but almost every freeliving animal and plant is food for something else. No matter how tiny, large, swift, camouflaged, armored, spiny, or poisonous it is, some other animal can find it, catch it, overcome its defenses, and consume it. As consumers, animals in general have an almost infinite adaptation for organic food, not only in their jaws and dentition but in their food detection techniques and digestive apparatus. For example, the mouths of fish of different species are adapted to cut, grasp, suck, stab, rasp, scrape, dig, crush, and filter. Some invertebrates can bore through wood and the hardest of molluskan shells. The adaptations are usually consistent from detection of prey through capture and mastication to digestion. An animal that feeds by filtering plant materials will probably have a small gullet, enzymes suitable for digesting carbohydrates, and a large intestinal surface. An animal that catches and gulps large fish will probably have a large gullet, a large and unusually acid stomach, and a small intestinal surface.

Usually the larger and more highly organized animals feed on smaller and more primitive organisms, but there are many exceptions. Some fish can swallow animals larger than themselves; some jellyfish eat fish; some starfish eat mollusks; some fish eat birds or mammals.

At the start of life all animals differ from adults in their capability to find, capture, and ingest food. Most animals have completely different kinds of diets at different stages in their lives.

Aquatic animals find their food by using a variety of senses, particularly smell and sight. Many predators are assisted by the characteristic odors of their prey. It is well known that many fish and crustaceans are attracted by either the body fluids of injured animals or pieces of flesh. Having approached a prospective prey, the foraging animals must depend largely on vision for the final attack. Other sensory systems used by certain animals include methods of detecting noise, movement of the water, changes in electrical resistance, and echo ranging.

The importance of vision in the feeding of many animals is indicated by the prevalence of circadian rhythms in feeding. Many foragers feed more vigorously at dawn and evening than at night and noon, although they may do so as a result of the behavior of the prey.

Most animals in temperate climates have seasonal rhythms in their feeding that are caused also by combinations of circumstances. Commonly the amount of food needed is greater in warm water than in cold because at lower temperatures digestion and metabolism are retarded. At temperatures near freezing some species cease to feed. The spawning season for some animals is a period of fasting. Perhaps the most common cause is a seasonal cycle in food abundance. Many animals feed intensively during the plankton blooms, the emergence of certain insects, the spawning periods of crabs and fish, or other seasonal phenomena, each of which may last only a week or two.

No simple classification of feeding systems is completely satisfactory. Some animals are herbivorous, feeding entirely on plants; some are carnivorous, feeding entirely on animals; but many are omnivorous, feeding on both animals and plants either at the same time or at different times in their lives. The spectrum of habit ranges from the stationary filterer to the speedy forager and includes the parasite.

4.5.2 DETERMINING THE DIET OF AQUATIC ANIMALS

Because it is rarely possible to observe an aquatic animal feeding under natural conditions, its habitual food or diet is determined by an examination of the stomach contents of a suitable sample of animals. Usually stomachs are easy to obtain from fishermen or from animals collected for other purposes. They are usually preserved for later examination in a laboratory. There it is a straightforward though not often an easy task to identify each organism in the stomach and either weigh it or determine its volume.

More difficult is the task of analyzing and interpreting the data. Complications arise because animals feed irregularly, they sometimes regurgitate the most recently eaten items during the process of capture, food items digest at varying rates, and identification of digested food items may depend on certain teeth or skeletal remains.

The objectives of dietary studies are to understand how animals live and grow, what foods may influence their abundance and distribution, and the relative quality of feeding conditions. Usually such studies are part of life history studies of such matters as distribution, migrations, growth, and reproduction.

Despite the difficulties of interpreting the data on number and weight (if volume is taken, it is assumed to have a specific gravity of 1) of food items in stomachs, a great deal can be learned. The counts of different food items may be compared directly with each other, but in this method extra emphasis is given to items with parts highly resistant to digestion and to the tiny items, e.g., a thousand crab larvae may equal the food value of only a single fish. A much better method is to express the frequency of occurrence of each item among the stomachs examined by showing the percentage of stomachs containing it (Table 4.1). Such a percentage shows the number of individuals in the sample that ate the food item and hence a rough index of the availability of the food. A still better method is to use the volume of each item since the emphasis given to items with durable parts is less. The volume of each item may be reported either as a percentage of the aggregate volume of all stomachs or as an average of the percentage in each stomach—a method that gives each stomach equal weight regardless of the volume of contents.

When a representative sample of the food available is obtained along with the stomachs, it is possible to estimate a *selectivity coefficient* or *forage ratio* for each food item. This ratio is taken by dividing the percentage in the stomach by the percentage in the environment. A value higher than 1 in-

TABLE 4.1
EXAMPLE OF A FOOD ANALYSIS[a]

Food organisms[b]	Number of organisms	Percent of stomachs in which occurred	Percent of total volume
Crustacea	85,140	66.9	24.8
Squid	3642	55.4	26.2
Fish	5333	70.4	46.7
Flying fishes	23	1.9	3.3
Jack, *Decapterus*	46	2.2	8.4
Pomfret, *Collybus*	449	13.8	3.4
Tuna, *Katsuwonus*	19	1.5	5.1
Unidentified fish	2439	48.2	8.0

[a]Food organisms found in the stomachs of 1097 yellowfin tuna captured in the central Pacific, 1950 and 1951. After: J. W. Reintjes and J. E. King (1953). Food of the yellowfin tuna in the central Pacific *U.S., Fish Wildl. Serv., Fish. Bull.* **54**(81), 91–110.

[b]Table is greatly abbreviated—all organisms were identified as accurately as possible and many more were listed.

dicates selection, lower than 1 avoidance. The difficulty of obtaining a satisfactory sample of the food in the environment with either nets or grabs limits the determination of this coefficient to animals feeding entirely on either small plankton or small bottom invertebrates.

The total volume of all items in a stomach expressed as a percentage of the weight of the animal gives an *index of fullness*. The index is useful for comparing the relative feeding activities at different times and places.

Special studies of captive animals in which the total weight of food eaten is related to the gain in weight of the animal yield an estimate of the *nutritional coefficient* or *conversion ratio*. The nutritional coefficient is the number of units of food required to produce a unit gain in weight on a net weight basis. It varies with size and activity of the animal and the kind of food. For fish, the value varies from 2 to 40 and a value of about 10 is common. Refined coefficients with various names and based on the dry weight of diet components (i.e., proteins) are also used.

4.5.3 EXAMPLES OF FOOD AND FEEDING HABITS

Stationary Bottom Filterer; the Oyster. The sessile animals of the bottom cannot forage but must filter their food from the water as it flows by. They increase the efficiency of the operation by increasing the size of the filter, by pumping water through an enclosed filter, and by sorting the nourishing from the nonnourishing particles and the harmless from the harmful.

Adult oysters pump a surprisingly large amount of water over their gills which serve not only as respiratory organs but also as filters and sorters. Large American oysters, *Crassostrea virginica*, have been found to pump amounts varying from 10 to 20 liters/hour when active and up to several hundred liters/day although their periods of activity are highly variable. Water is pumped over the gills through the open shell by the action of the cilia which cover the gills and much of the body. The values of the shell are not moved to pump the water. Water is drawn into the ventral region, passed by diverse routes over the gills, and exhaled posteriorly (Fig. 4.7). The particles in the water become entangled in the mucous on the gill surface and are pushed toward the mouth. Either along the way or at the labial palps near the mouth they are sorted; the food is passed into the mouth; the waste is pushed out as pseudofeces. This filter system will remove particles from the water as small as 2 μ.

The food of oysters is not satisfactorily known because a large proportion of it is the naked and minute nannoplankton that is too delicate to examine easily even before ingestion by an animal. Despite disagreement about details of their food it is generally known that the oysters, like most other bivalve mollusks, consume mostly phytoplankton (see Chapter 5 on basic ecology), and hence they are relatively efficient in transforming the basic product of the water into food which can be consumed by man.

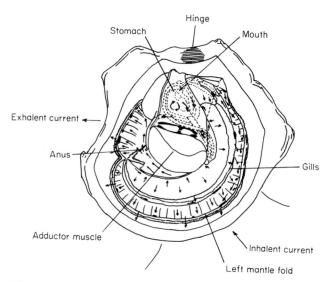

Fig. 4.7.   Ciliary water currents during feeding of the European oyster, *Ostrea edulis*. Right shell and mantle removed. [Redrawn from C. M. Yonge (1926). Structure and physiology of the organs of feeding and digestion in *Ostrea edulis. J. Mar. Biol. Ass.* **14**, 295–386.

Carnivorous Bottom Forager; the Cod.   The cod might well be selected as a bottom-dwelling fish with the least specific food habits. It has a mouth which is about average in size and power. Its jaws and the roof of its mouth are equipped with numerous rows of small, pointed teeth suitable for holding animals but not for tearing or crushing. Its gill rakers and gut have no unusual features to suggest other than an ordinary food habit.

As soon as the larval cod consumes its stored food it must feed regularly on particles of the proper size. At first the food is small phytoplankton, but soon the pelagic larva is large enough to eat tiny crustaceans such as copepods, barnacle larvae, and crab larvae. As it grows it becomes a bottom forager and turns to larger crustaceans such a shrimp or euphausids, to miscellaneous bottom invertebrates such as tunicates, brittle stars, sea cucumbers, squid, clams, or mussels, and to fish in great variety. It eats almost any small fish, including young cod. After the cod reaches a length of about 50 cm, fish is the predominant item in the diet.

Such a varied diet indicates a ready adaptability to changes in the abundance of food. Apparently cod can utilize whatever animal food is caught easily and are found commonly to eat different diets in different places at different times of the year.

This readily changeable food habit is typical of many carnivorous foraging fish with ordinary dentition in either fresh waters or the sea and for either pelagic or bottom-dwelling species. Most of them tend to feed more on invertebrates (insects in fresh water) than fish when small; and more on fish than invertebrates when larger.

Giant Filter Feeders; the Whalebone Whales.   As the size of an animal's food becomes smaller relative to the size of the animal, more energy must be expended to find and capture the food relative to the energy provided if the food items must be taken singly. Food particles each weighing less than about 0.1% of an animal's daily ration may not supply enough energy unless they can be taken in quantity by some kind of a filter system.

The largest animal filter system is possessed by the whalebone whales which have up to 300 baleen plates on either side of their mouths. These plates may be up to 4 m long in right whales and up to 1 m long in rorquals. They are arranged about 1 cm apart in a row around the upper jaw, and are fringed on the inner edge with hairs. These hairs form a matlike strainer that prevents particles larger than about 2 mm from getting out. The whales operate this apparatus either by swimming along with their mouths open and occasionally swallowing, or by striking at a dense aggregation of food. The food of the whalebone whales in the Antarctic is mostly krill, a euphausid crustacean that varies in length from 2 to 5 cm. In the Arctic, however, krill is not as abundant and their food consists of large copepods, other crustaceans such as crab larvae, or schooling fish such as smelt, mackerel and herring.

It is interesting that a large part of the whale's food is also comprised of filter feeders. The krill, like most of the small crustaceans, is equipped with rows of hairlike projections on its forward legs that strain and move particles toward its mouth. It feeds principally on diatoms larger than about 0.04 mm. The fish of the sizes and species taken by the whales are partial filter feeders. They feed principally on small crustaceans but may include larger colonial phytoplankton in their diet, especially when small.

4.6 Reproduction and Early Development

Most aquatic organisms spend much of their lives and energies reproducing. After a brief juvenile stage, they develop sperm or eggs, spawn, recover, and repeat the process in a cycle that continues until senility or death. The sperm and eggs are usually produced in such quantities that they represent a large expenditure of energy. The spawning process may be preceded by migrations or nest building and followed by care of the young. The postspawning recovery is usually coincident with or immediately followed by preparation for the next spawning.

The eggs and larvae start a life that is completely different from the life of the adults. The young of all sessile animals are mobile for a time. Commonly the number of young is related to their fragility. The pups, or calves of marine mammals number one or two at a time and are usually able soon after birth to feed and live much like the adult. At the other extreme the eggs produced by an oyster or a shrimp number in the millions and hatch into feebly swimming larvae, completely unlike the adults. Most such larvae drift with the currents in which they are easy prey for other animals.

The cycle of breeding activity is closely correlated with feeding activity, and these two activities account for most of the migrations and aggregations of the aquatic animal resources. Both the process of developing sperm or eggs and the process of recovering from spawning require extra energy and good feeding conditions. Some animals store large amounts of energy, in addition to that in the gonads, which is used either while migrating to spawn or while fasting during spawning.

The fishery scientist is commonly concerned with many aspects of the reproductive process. With wild animals he needs to understand the reproductive process either to catch them easily or to protect them if they are unduly vulnerable, or to explain the great fluctuations in abundance due to failure of the young to survive. With domesticated animals he will try to control the environment to make the process as efficient as possible. He will be concerned with preventing waste either of animals in poor condition as a result of breeding activities or of young being cared for by their parents.

4.6.1 MATURITY

The cold-blooded animals with which we are concerned do not attain a definitive size or social behavior that may be used to define maturity as in birds or mammals. The period of life in which sexual maturity is attained varies greatly; a few species are mature at birth, others mature only once near the end of their lives. Most of the higher aquatic animals mature at an age equivalent to between one-fifth and one-third of their maximum life-span. Within a single species the better-fed individuals living in warmer waters are the first to mature and frequently males will mature at a younger age than females.

Some differences in the definition of *maturity* will be found among authors; some use the term to designate animals at all times after they are ready for their first spawning, whereas others restrict the definition to animals during the time that they are actually spawning. In this volume the term in the first sense is used.

The onset of maturation is indicated by the rapid development of the gonads. Among juvenile fish the gonads are slender strands of tissue in the dorsal part of the abdominal cavity. As the gonads increase in size the individual eggs in the ovary become visible to the naked eye and the testes change to a pale reddish color. When the gonads attain full size they seem almost to fill the body cavity; ovaries commonly make up between 10 and

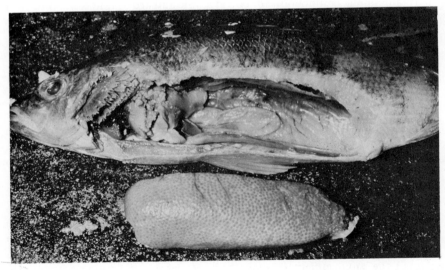

Fig. 4.8.   The ovaries of a mature yellow perch, *Perca flavescens*, fill almost all of the body cavity. (Photo by Professor Lynwood S. Smith, College of Fisheries, University of Washington, Seattle, Washington.)

25% of body weight (Fig. 4.8), testes between 5 and 10%. At this stage the fish are considered to be *ripe* and will soon spawn naturally. Soon after the gonads reach maximum size in some species such as salmonids the ova and the sperm (called *milt*) can be expressed from the body by light pressure on the abdomen and fertilization accomplished artificially.

4.6.2 FECUNDITY

Fecundity among egg-laying animals is the number of eggs being readied for the next spawning by a female. *Relative fecundity*, the number of eggs per unit of weight, is commonly used as an index of fecundity. *Total fecundity* is the number of eggs laid during the lifetime of the female.

Fecundity is more or less inversely related to the size of eggs and to the care given to eggs and larvae. Some sharks may have eggs between 6 and 9 cm long and only about 12 at a time. Scarcely more than this number of eggs are produced by some of the tropical fishes, but they are tiny and incubate either in their mouths, in special pouches, or in nests. Near the other extreme is the case of the animals that broadcast tiny, fertilized eggs to drift with the plankton. Most of the important marine fishes have this habit; many of the larger cods or flounders produce more than 1,000,000 eggs at one spawning. The largest number of eggs is produced by the sessile invertebrates, such as oysters, which may produce tens of millions of eggs at one spawning.

Fecundity is related to the hazards faced by the eggs, larvae, and young. Most animals maintain a reasonably constant population size. In bisexual animals, with a sex ratio of one to one, each female may be expected to produce enough eggs during her entire life to result in an average of two more adults. The species that have survived the process of evolution have been only those that have produced enough eggs and given them enough care to maintain their numbers. (See Section 5.1.5.)

One characteristic of the more fecund species is especially important to the fisheries: they fluctuate more in abundance than the less fecund species. Apparently the survival from many spawnings is very poor, whereas the average survival of two adults is maintained by an occasional highly successful spawning. Among fish that spawn annually, the progeny of such a spawning is called "a dominant year class," among mollusks, "a successful set of spawn." Frequently the fishery scientist is asked to explain whether such fluctuations in abundance are due to natural causes or to the effects of fishing. Such species also offer him the tantalizing challenge of discovering and controlling the limiting factors in order to obtain a successful spawning every year.

4.6.3 SECONDARY SEXUAL CHARACTERISTICS

Most mature fish and crustaceans develop external sexual differences in addition to the primary sexual characteristics, the gonads. These may help either man or prospective mates to recognize the sexes; they are called *secondary sexual characteristics* and may be either accessory to spawning or not. The accessory structures include the intromittent organs of some male crustaceans and fish and certain fin or mouth structures that may be used in the care of either eggs or young. Other sexual differences (which are more common among freshwater fishes) include differences in size, shape of body, shape of fins, color, and presence of tubercles. Color differences between the sexes and special nuptial colors for both sexes are most striking among the salmonids and many small freshwater tropical fish, which are prized by aquariists. Perhaps the least sexually differentiated fish are the marine species that broadcast their eggs in the water. The sex of most of the cods, herrings, tunas, and mackerels cannot be recognized outside the spawning season except by dissection.

4.6.4 SEX REVERSAL

In most vertebrates, sex is determined at the time of fertilization and is fixed for life. However, in some invertebrate resource animals sex may change regularly or erratically, but rarely are these animals true hermaphrodites.

In many species of oysters, the oyster matures first as a male, slowly becomes a female, quickly becomes a male again, and then continues to alternate in sex throughout life. In colder climates oysters may change sex once a year, in warmer climates several times.

A different schedule occurs in at least some species of *Pandalus*, the northern shrimp. These shrimp mature first as males, remain males for a year or two, become females, and remain so for the balance of their life. A consequence of this habit is that a fishery seeking the larger individuals may catch a preponderance of females and deplete seriously the female population.

4.6.5 TIME OF SPAWNING

Among aquatic animals spawning occurs at a time after the adults have been able to feed well enough to have extra energy for the developing gonads and when they can escape predators. Spawning and incubation must also precede a period of favorable food production sufficient to sustain the young. The timing for each race of animals has been determined by evolutionary adaptation; those animals that survived have originated from spawn released under favorable conditions.

For many aquatic animals in the temperate zones spawning appears to occur in advance of the spring blooms of plankton by an interval about equal to the time required for the embryo and larva to consume the energy stored in the egg. Usually the spawning time is in the late winter or spring, but salmon in the northern hemisphere require between 3 and 9 months to pass through the egg and prolarval stages and commonly spawn between July and December. Much variability exists, however, even within the same species. This is especially prevalent for species in tropical and subtropical areas where environment is relatively uniform throughout the year.

Regardless of the variability among animals, some kind of schedule is followed by each race. The schedules are controlled in most aquatic animals at least in part by temperature and length of daylight. These factors may operate separately or jointly. In experiments spring-spawning animals spawned sooner when either temperature or the period of daylight was increased artificially; autumn-spawning animals spawned sooner when either temperature or the period of daylight was decreased. Other factors that may operate are tides, salinity, and floods.

The effects of such exogenous factors are, however, always subject to the restraints of the animal's own endogenous cycle. For example, an unusually warm spring may hasten spawning for an animal, but only if the gonads have developed adequately already—a process that may have started many months before.

4.6.6 SPAWNING BEHAVIOR

In addition to a schedule, every bisexual aquatic animal has a spawning ritual. Each must locate and recognize a member of the opposite sex of the same species, or recognize the other's sexual products, then release its own ova or sperm at the right place and time. The rituals and schedules are as varied as the number of races of animals; indeed they are important aspects of the adaptation of the races to their environment. For each race of each species, however, they are relatively invariable. The procedures and sequences must be followed if the spawning is to be successful.

A few species engage in courtship during the later stages of gonad development. This may include pairing off, displays of colored fins, and aggressive behavior between members of the same sex. Such behavior is more common among bottom-dwelling freshwater fishes than among the pelagic marine species, many of which congregate in large spawning shoals.

Commonly, fertilization among fish occurs outside the body, in the water. It is accomplished by the nearly simultaneous release of sperm and ova close together: usually the ovum is fertilized within seconds of being laid. Fertilization by the sessile mollusks also occurs in the water. Both sperm and ova are released to drift with the currents and find each other, either in the mantle cavity or free in the sea. (See Section 4.1.2.)

Fertilization takes place internally in many crustaceans and in a fair number of fish. The male has an intromittent organ which is used to place the sperm in the female's body. Fertilization may be effected immediately, or the sperm may be stored for a period. Usually the energy for the developing embryo and larva is supplied by the yolk of the egg, whether fertilization is internal or external. In a few species of fish, the female's body will provide additional energy through some kind of a placentalike structure.

Some freshwater fish and bottom-dwelling marine fish care for their young for a time. Most nest builders guard their nests and young; some carry the incubating eggs and larvae about in their mouths or in special pouches. Many crustaceans carry their eggs and smallest larvae attached to abdominal appendages on the female. Marine mammals usually bear one or two living young after a gestation period of as much as 15 months.

4.6.7 INCUBATION

When one sperm enters the ovum, usually other sperm are excluded; incubation starts immediately and continues until the egg hatches. The single-called blastodisc divides and differentiates to form an embryo. During this exceedingly complicated and delicate process the energy of the yolk is used according to the coded information in the genetic material to form the new animal, and energy is obtained and carbon dioxide and other wastes are eliminated through the egg membrane.

The incubating eggs of fish usually develop a prominent pair of dark spots at about the middle of the incubation period; these are the eyes of the embryo. These are called *eyed eggs* by fish culturists. The development of the eyes in trout and salmon eggs coincides with the end of a period of extreme sensitivity to shock that starts soon after fertilization.

The period of incubation or gestation varies among species of fish from a few hours to more than a year; the longer periods being required by fish that spawn in the autumn at high latitudes. The eggs of many of the common marine food fish hatch withing 1–2 weeks. The period for each species varies inversely according to temperature within the tolerable range of temperatures. For example, trout eggs will tolerate incubation temperatures between 2° and 15° C. Within this range, the eggs of a particular race require about the same number of degree days. When the race requires about 400 degree days, the egg hatches in about 40 days at 10° C, in about 80 days at 5° C.

4.6.8 LARVAL DEVELOPMENT

The moment of hatching is the beginning of a hazardous period in the life of the aquatic animal. It marks no obvious change in the structure of the larval fish, which merely is freed from its confinement in the shell, but it creates the need for a profound physiological and behavioral change. The tiny animal must shift its respiratory and excretory systems from an exchange through the egg fluids amd membranes to an exchange directly with the water through its own gills, skin, and kidneys. It must maintain its own osmotic balance without the help of the egg membrane. It becomes a swimming object and thus is more likely to attract the attention of predators than the motionless egg. It must find, catch, and digest its own food as soon as its supply of yolk is used up.

The larva begins a new life that is not only entirely different from its life as an embryo but also entirely different from the life it will lead as a juvenile or adult. It probably does not resemble the adult: it does not eat the same food, and it has a much higher metabolic rate. It leads a completely different life—probably not even in the same place as the adult. Furthermore, its life is far from static because, relatively, it is growing faster than it ever will grow later in life, changing in structure and habit every few hours or days as its organs develop.

Larval development continues until the animal reaches the juvenile stage, when it more or less resembles the adult. A larval fish, while still using its stored yolk, is called either a *prolarva* or a *sac fry*. After it has absorbed the yolk it is called either a *postlarva* or an *advanced fry*. The larval forms of invertebrates are much more varied and numerous. Usually bivalve mollusks have such an indistinctly structured egg shell that no real hatching occurs. The embryo develops while drifting in the water into a feebly swimming larva called either a *trochosphere* or *trochophore*, from the Greek "trochos," wheel. The name refers to the wheellike rings of cilia. A second stage and much better organized swimming larva is the *veliger*, from the Latin "velum," veil, and "genere," to carry. Crustacea usually hatch first into a free-swimming *nauplius*, an unsegmented, eggshaped larva with three pairs of appendages. A juvenile stage is attained after the next moult in some species, not until after several intervening stages in others. Two of the common intervening larval forms are called *zoea* and *megalops*; others are named after other adult crustaceans that the larvae happen to resemble.

4.6.9 EXAMPLES OF REPRODUCTION AND EARLY DEVELOPMENT

The Oyster.   The ova are fertilized within the mantle cavity of the female *Ostrea*; in the open sea in *Crassostrea*. In either location development proceeds for a few days until a ciliated velum has developed. The velum propels the animal and sweeps food into its mouth. At this stage it is called a *trochosphere*. It is about 0.2 mm long and capable of feeding on plant cells no larger than about 0.01 mm. It soon becomes a better organized, veliger larva; and after between 1 and $2\frac{1}{2}$ weeks, depending on temperature, it develops a foot and begins to crawl. It has also developed a shell and become heavier. It can search on the bottom for a place to attach itself, since it can still

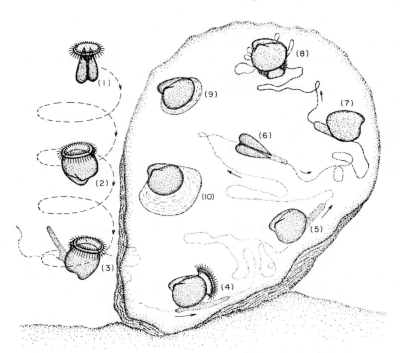

Fig. 4.9.   The setting process in the American oyster *Crassostrea virginica*. (1) and (2), Swimming larvae with protruded velum; (3) and (4), searching phase with protruded foot; (5) to (7), crawling phase; (8), fixation to substratum; (9) and (10), spat 1 and 2 days old. The size of the larvae and spat relative to the old oyster shell is greatly exaggerated, the former being about 0.3 mm long at stage (1). [Source: H. F. Prytherch (1934). The role of copper in the setting, metamorphosis, and distribution of the American oyster *Ostrea virginica. Ecol. Monog.* **4**(1), 47–107.]

swim. After more or less exploration it *sets* by attaching itself on its left valve by means of an adhesive cement. Once it is attached, the velum and foot disappear within 2–3 days, and the shell rapidly enlarges. The *spat* metamorphoses to an animal generally resembling the adult when the shell measures only about 1.5–2.0 mm in diameter (Fig. 4.9).

This process and the factors controlling it have been studied intensively because it is fundamental to oyster culture. An important factor is temperature; after oysters are mature and in water near 20° C they may be induced to spawn. They commonly spawn in several pulses during the summer.

The oyster grower tries to put spat collectors in the water at just the right time to collect the naturally spawned spat and avoid the larvae of other animals, which foul the collectors if they are submerged continuously. Sometimes government agencies provide a forecast service of setting prospects for oyster growers based on temperature and quantitative samples of oyster larvae in the water.

Penaeid Shrimp. Most of the edible shrimps, crabs, and lobsters either live in the sea or migrate between the sea and the estuaries. Most of them produce weakly swimming pelagic larvae. The larvae may drift with the ocean currents for a few days or a few months depending on species. Most species spawn at a single season of the year in a place in the ocean from which the young will drift to suitable nursery grounds in shallow water.

The penaeid shrimps, the common commercial shrimps of the tropics and subtropics, generally make a migration from shallow water to deeper water to spawn. In most species the male deposits the sperm capsule on the abdomen of the female to fertilize the ova when released by the female.

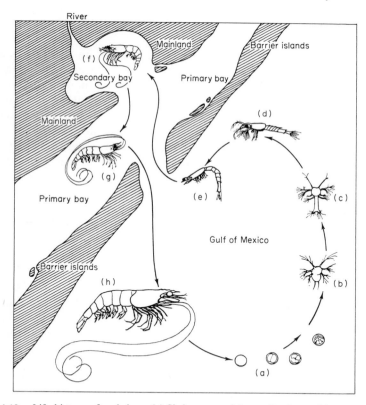

Fig. 4.10. Life history of a shrimp. (a) Shrimp eggs; (b) nauplius larva; (c) protozoea; (d) mysis; (e) postmysis; (f) juvenile shrimp; (g) adolescent shrimp; (h) mature adult shrimp (not drawn to comparable scale). [Source: A. W. Moffett (1967). "The Shrimp Fishery in Texas," Bull. No. 50. Texas Parks and Wildl. Dep., Austin, Texas.]

Several hundred-thousand semibuoyant eggs are laid by each female. These hatch in about 1 day and then go through a series of about 10 moults as they grow gradually and transform into tiny shrimplike animals. These stages drift with the currents back toward the estuaries, which are the nursery grounds. Here, when the water is warm, the young shrimp grow rapidly, and after a few months become adults. The adults start back to sea to spawn; usually they spawn only once during their lifetime (Fig. 4.10).

Atlantic Mackerel, Scomber scombrus.   Many of the marine food fishes produce large numbers of small eggs that they broadcast in the sea to drift with the currents. Many of these species also experience spectacular changes in abundance due to variation in the survival of the young. The Atlantic mackerel, an important food fish of Western Europe and Eastern North America, is one of these species selected as an example because of the extensive studies of the early history reported by Sette (1943), from which this account has been taken.

The mackerel are schooling fish. They reach a maximum length of about 55 cm but mature generally at a length of about 35 cm and an age of 2 years. They live in the open sea, mostly over the continental shelf, but are not dependent either on the coastline or on the bottom at any stage in their lives. They winter in depths ranging from 100 to 200 m along the outer part of the continental shelf and move toward the surface, toward the shore, and toward the northpole during the spring on a feeding and spawning migration. They are almost always found in water temperatures between 7° and 20°C.

The fishery is pursued during the late spring, summer, and early autumn when the mackerel are nearer the shore and the surface. It has been an important fishery for centuries in Europe and since early colonial days in North America. Its changing abundance has plagued and puzzled the fishermen, however, who find them in enormous numbers in some years and in scant supply in other years. North American production has varied from 6000 to 106,000 tons annually.

Approximate data on the reproductive process are as follows: Spawning begins in mid-April in the southern part of the North American range, in mid-July in the northern part (Fig. 4.11), but continues for only 3–4 weeks in any one locality. An average female produces 400,000 ova. The fertilized egg is 1.2 mm in diameter, drifts in the upper 10 m layer of the sea, and hatches in 4 days. The newly hatched, feebly swimming, larva is 3.2 mm long. It absorbs the yolk sac in 5 days, develops fins in 26 more days when 10 mm long, and then reaches the adult stage at a length of 50 mm 40 days later (Fig. 4.12).

These findings were established or verified by Sette as a preliminary to determining the rates of growth and mortality of the eggs and larvae. This

Fig. 4.11. Mackerel spawning areas along the Atlantic Coast of North America. Relative intensity is indicated by the average number of eggs caught in plankton nets. [Source: Sette (1943).]

he did from an extensive series of quantitative net samples taken repeatedly over the entire range of the United States stock and throughout the period of spawning and larval development in 1932. After identifying and sorting the mackerel from the numerous other species in the catches, he was able to find *homologous* groups, i.e., eggs fertilized in a brief period in a limited area that could be identified in later catches when further developed. After determining the change in length of each homologous group, he was able to determine the rate of growth. Then, from the numbers of mackerel taken in the same samples, he was able to estimate the actual numbers of eggs and larvae in the sea at different stages in their development.

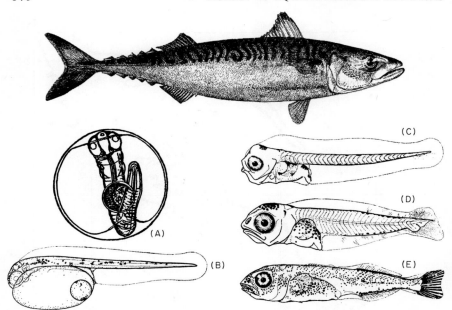

Fig. 4.12. Mackerel, *Scomber scombrus*. (A) Eggs; (B)–(E) larvae 3.5 mm, 4.6 mm, 7.8 mm, and 15 mm in length, respectively; (F) adult, 40 cm. [Source: H. B. Bigelow and W. C. Schroeder (1953). Fishes of the Gulf of Maine. *U.S., Fish Wildl. Serv., Fish. Bull.* **53**(74), 318.]

The rate of larval growth was found to be logarithmic; it proceeded at about 4%/day in length, 12.5%/day in weight! The number of eggs laid was estimated to have been about 64×10^{12} by a spawning population of about 1×10^8 individuals. These died during most of the developmental period at rates of 10–14%/day except as the larvae were developing fins, when the rate was 30–45%/day! Only between 1 and 10 mackerel reached a length of 50 mm out of each million eggs spawned!

These observations were the record of the failure of a year class due probably to two possible causes: the failure of the newly hatched young to find enough of the special food they needed, or the drift of the young with unusual currents away from their accustomed nursery areas. Sette concluded, on the basis of circumstantial evidence, that the second factor was probably the predominant cause of the failure of the 1932 year class although the scarcity of plankton may have been a contributing factor.

Sockeye salmon, Oncorhynchus nerka. The reproduction of salmon, trout, and chars has been studied more extensively than that of almost any other fish because it is spectacular and easy to observe, and it is easy for man to intercede for the purpose of rearing the young artificially. Almost all of these fish migrate from feeding to spawning grounds where they become brightly colored and develop sexual differences. The female digs a pocket, or redd, in the gravel in which she lays the eggs and covers them after fertilization by the male. The eggs hatch and the larvae consume the stored yolk while the eggs are protected by the gravel; then the larvae emerge to begin feeding. As an example of a species with this type of reproduction the sockeye salmon was chosen because it is typical of the group with respect to its egg laying and larval development, but more spectacular than most in its migrations and nuptial colors.

The sockeye salmon spawns only once—at the end of a life ranging usually from 4 to 6 years in length and migrations of thousands of kilometers in the ocean. It begins to migrate from the high seas toward its home stream in late spring, traveling at speeds of between 30 and 50 km/day through the open sea, to its estuary, and up its river to its lake system, which is its nuptial and nursery area. Its gonads have begun to develop rapidly while it is still at sea and it stops feeding as it approaches the estuary. Soon after entering fresh water it begins to change color from bluish silver to bright red in the body and green in the head. It finds its way through

Fig. 4.13. A pair of spawning chum salmon *Oncorhynchus keta* release a cloud of sperm. Their spawning behavior is almost identical to that of sockeye salmon. (Photo by Asko H. Hamalainen, University of Washington, Seattle, Washington.)

the lake to its spawning beach or stream and spawns either in late summer or in autumn.

The female usually digs a redd, then mates with an attending male (Fig. 4.13), and covers the eggs with gravel a few minutes later. She usually deposits her eggs in four to six clusters of about 500 eggs each. The preferred spawning site is in a pool at the head of a riffle or in a riffle through which water flows, with gravel small enough and loose enough to be moved around by the female and with little fine sand or silt that would prevent circulation of the water or the emergence of the larva.

The eggs usually hatch either in late autumn or in early winter; the larvae remain in the gravel subsisting on their yolk sac (Fig. 4.14) until early spring when they wriggle up through the interstices of the gravel. They emerge and seek food and shelter along stream banks and lake beaches for several weeks, then move out to the open water of the lakes where they lead a pelagic

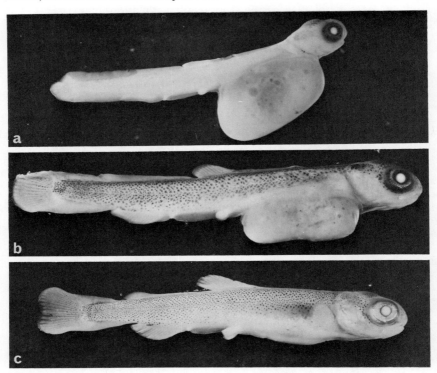

Fig. 4.14. Young sockeye salmon, *Oncorhynchus nerka*, soon after hatching (a), middle of sac fry period (b), and beginning of postlarval period (c). The latter stage is the time of emergence from the gravel. The mean lengths are about 17, 22, and 25 mm, respectively. (Photos by James C. Olsen, Fisheries Research Institute, University of Washington, Seattle, Washington.)

life. Their life in the lake continues until the following spring, or even until the second following spring, when at a length of between 5 and 10 cm they obey their urge to migrate down the river to the sea.

4.7 Age and Growth

Growth is commonly considered to be a gradual increase with time in size or mass or some kind of a living unit. It may be applied to a part of an organism, to a whole organism, or to a population of organisms. It is complicated by the differentiation of parts of organisms, which may grow at different rates, even negative rates (degrowth), and by the death of parts of populations. It is further complicated because the use of energy for growth is not greatly different from the use of energy for maintenance of the unit, for the repair of wear and tear, or for the excretion of products. A more exact definition is "Growth is the addition of material to that which is already organized into a living pattern." In terms of energy equivalents it may be expressed as

$$\Delta W / \Delta t = R - T$$

in which ΔW is the growth, R the total energy obtained from the rations, and T the total energy expended in metabolism all during the unit of time Δt. This section deals with the growth of aquatic organisms, and a later section deals with the growth of aquatic populations.

Few topics are of more fundamental importance to the fishery scientist than the age and growth of aquatic life. Such questions as these are constantly recurring: How old is this animal? How much will it weigh next year? At what age does it mature? At what rate does it grow at age x? When is it the fattest? When will the progeny of these parents be large enough to catch? Answers to such questions are basic to understanding the life of individuals as well as a necessary preliminary to understanding the fluctuations in resource populations.

Two kinds of growth are found among the living aquatic resources: the stepwise growth of the crustaceans, and the "continuous" growth of the mollusks and the vertebrates. The difference is not really too important for most purposes, however, because much of the growth of crustaceans at moulting is due to the addition of water to energy already stored and much of the so-called continuous growth of mollusks and invertebrates varies cyclically in rate with the seasons and schedules of activity.

The growth of almost all of the aquatic resource animals is asymptotic, that is, each species in each environment has a characteristic ultimate size which it approaches by growth throughout life. This type of growth is common to all cold-blooded vertebrates and, surprisingly, to aquatic mammals as well which do not reach maximum size at about the age of maturity as many of the terrestrial mammals do. An important characteristic of this type of growth is the relative fixation of the typical body proportions early in life rather than during adolescence and maturity, as in the terrestrial mammals.

4.7.1 MAXIMUM AGE AND SIZE

Age and size of aquatic resource animals are more or less correlated. The oldest among the animals tend to be the largest, but there are many exceptions. The long-lived fishes tend also to be those with the following characteristics: they are phylogenetically primitive, sluggish in their movements, bottom- or shallow-water inhabitants; they have accessory respiratory devices; and they are adaptable to extreme fluctuations in oxygen concentration, temperature, and salinities. Examples are sturgeons, sharks, and carp. The short-lived fishes are those with opposite behavior and adaptability, including tunas, salmon, and capelin.

The maximum age of wild aquatic animals is not well known because those that die naturally are seldom seen and the evidence of age in the oldest of the animals is frequently obscure. The oldest fish on record (albeit a questionable one) is a wild sturgeon whose age was estimated from the rings in its fin rays as 152 years. Many fish have been kept for long periods in aquaria; the records go to a sturgeon kept in the Amsterdam Aquarium for 69 years and a carp kept in the Frankfurt Aquarium for 38 years. Numerous species of larger fishes have been kept for more than 20 years. Some invertebrates also may live a long time; northern lobster 50 years, king crab 20 years, and Pacific oyster 40 years. Apparently, such animals live longer than whales which start life as a relatively large infant, become sexually mature between 3 and 8 years, and survive for a maximum of about 40 years.

The largest living animal, the blue whale, now nearly extinct, is said to reach a length of about 30 m. One blue whale, 27 m long, was weighed carefully at a whaling station and found to weigh 119 tons.

The largest fish is the whale shark; it reaches a length of about 15 m. The largest bony fish is probably either the black marlin or the blue marlin; both exceed lengths of 4 m and weights of 700 kg.

The largest of the individuals of numerous species of fish caught by anglers are of special interest as trophies. The principal marine trophies are registered with the International Game Fish Association of For Lauderdale, Florida. The Association sets standards for the tackle that may be used, checks reports, and publishes annually lists of record fish for men and women and for various strengths of line. The all-tackle record fish prior to 1968 was a white shark, *Carcharodon carcharius*, of 1208 kg (2664 pounds); the largest bony fish a black marlin, *Istiompax marlina*, of 707 kg (1560 pounds). Records of North American freshwater fish are kept by *Field and Stream* magazine of New York. Regional and local records of trophies are also kept by many clubs and government fishery departments.

4.7.2 PATTERNS OF GROWTH

The growth of any animal is accompanied and influenced by many factors including both the endogenous events of its schedule of development from embryo through maturation to senility and the exogenous changes in its environment. Many of these factors operate independently in ways that are not well understood. Some influence the change in size of the whole animal, others the shape of the animal. Commonly, they must be studied empirically, i.e., by actual observation of the growth of a given species.

Endogenous Factors. In any organism a basic pattern of growth is inherited from the parents. Inheritance of rapid growth in trout was shown by breeding experiments in which rapidly growing adults were selected as parents in successive generations. The progeny of such selected trout in 1 year attained the length usually attained in either 3 or 4 years.

The growth and differentiation of animals is controlled by hormones. Their role in aquatic resource animals has not been as extensively studied as for some terrestrial mammals. The pituitary gland produces growth hormones; fish have been shown to respond when fed purified extracts of either fish or mammalian pituitaries.

Another factor that contributes to a slightly increased growth in fish with large eggs is the relative amount of yolk. Larvae from eggs with more yolk average slightly larger in size than those from eggs with a small amount of yolk and retain their size advantage for some time.

The larval period is characterized by rapid rates of growth and sudden changes in food and function, and is commonly also a period of rapid changes in body shape (Fig. 4.12).

After the animal begins to look like the adult the endogenous changes in growth are associated with the varying proportions of the energy input used for growth and other needs. The juvenile animal uses energy primarily for growth, secondarily for maintenance. Later energy is diverted to the maturation of the gonads and to preparation of the body for any fasting during spawning. After spawning, when the bulk of the gonads is gone, typically, the animal requires some time to regain its average body shape and replenish any depleted stores of energy. A cycle of varying partition of energy among growth, maintenance, and gonads continues until senility. At this time most of the energy input is used for maintenance and very little for growth.

Some species have sexual differences in growth. In many flatfishes the females are heavier for a given length, grow faster, and live longer than the

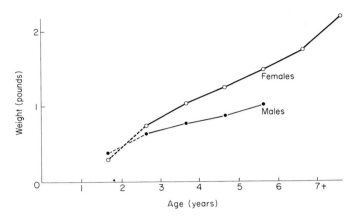

Fig. 4.15. Annual growth in weight of male and female yellowtail flounder, *Limanda fer-ruginea*, from Southern New England. Data are shown for the autumn, when effects of gonad development are minimal. [Source: W. F. Royce, R. J. Buller, and E. D. Premetz (1959). Decline of the yellowtail flounder (*Limanda ferruginea*) off New England. *U.S., Fish Wildl, Serv., Fish. Bull.* **146**(59), 169–267.]

males. An example is the yellowtail flounder (Fig. 4.15). In other species, such as the common dolphin, the male grows faster than the female. Such sexual differences occur also in some crustaceans; the male northern king crab is larger than the female.

Exogenous Factors.   The growth of any animal is influenced by a complex of environmental factors so that different stocks frequently grow at different rates (Fig. 4.16). Both the quantity and quality of food is probably the most important, but temperature too has a major effect at times. The combined effect of these two factors is usually marked in the fresh waters of temperate or arctic regions that freeze in winter. When the water is near 0°C both metabolic activity and growth are minimal (Fig. 4.17). The variation in food supply is not restricted, however, to the colder areas; most waters every-where have either blooms of plankton, swarms of insects, or schools of larvae, and these provide food supply for gorging and fattening (Fig. 4.18). One of the effects of the abundance of food on growth appears to be linked to the relationship between the energy cost of the food and the energy supplied. Small particles and hard to find food requires the expenditure of more energy for grazing or search, and therefore, reduces the proportion of energy input that is available for growth.

Growth may change when the animal is migrating, when extra energy is required for locomotion, when food is changing, or when the animal chooses not to feed. Some fish stop feeding entirely during the last part of a migration to the spawning ground.

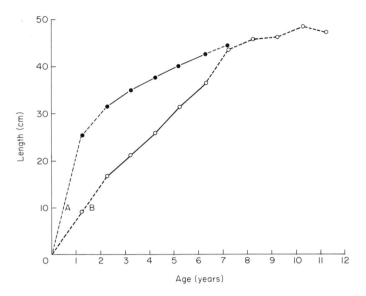

Fig. 4.16. Annual growth in length of female yellowtail flounder, *Limanda ferruginea*, from two stocks off eastern North America. Dashed lines are used where data are few. Curve A, Southern New England stock (Royce *et al.*, 1959); curve B, Nova Scotian stock (Scott, 1954). [Sources: W. F. Royce, R. J. Buller and E. D. Premetz (1959). Decline of the yellowtail flounder (*Limanda ferruginea*) off New England. *U.S., Fish Wildl Serv., Fish Bull.* **146**(59), 169–267. and D. M. Scott (1954). A comparative study of the yellowtail flounder from three Atlantic fishing areas. *J. Fish Res. Bd. Can.* **11**(3), 171–197.]

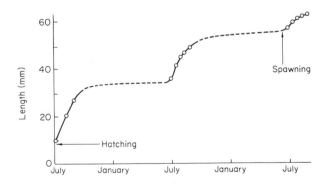

Fig. 4.17. Growth in length of threespine stickleback, *Gasterosteus aculeatus*, in a sub-arctic lake in Alaska. Dashed lines show presumed growth during winter. [From: D. W. Narver, (1966). Pelagial ecology and carrying capacity of sockeye salmon in the Chignik Lakes, Alaska. Ph.D. Thesis, University of Washington, Seattle, Washington.]

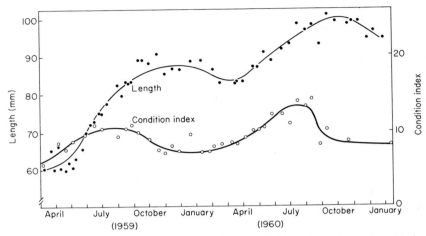

Fig. 4.18. Growth and condition index of oysters held on trays in Willapa Harbor, Washington. Curves are fitted by eye. Decreases in length are caused by breakage of the shell edges during handling. Condition index is dry weight of meat divided by volume of shell cavity × 100. [Data from K. K. Chew (1961). The growth of a population of Pacific oysters (*Crassostrea gigas*) when transplanted to three different areas in the state of Washington Ph.D. Thesis, University of Washington, Seattle Washington.]

The amount of living space per animal may affect growth in ways that are not well understood. Crowded animals may suffer from an accumulation of body wastes, a low supply of oxygen, or a small opportunity to get food. Any one of these factors would depress the rate of growth. Even when an abundance of food and water is supplied, still the animals may not grow as rapidly as less crowded animals. At the other extreme, animals either in isolation or in low concentrations may not grow as rapidly either. An optimum degree of crowding may enhance growth, presumably as a result of obscure social or chemical factors.

Length–Weight Relationship.   Every animal grows in both length and weight; the relationship between these is of both theoretical and practical importance. Comparison of the relationship indicates changes in body shape or in the condition of the animal. Further, it is necessary for a fishery scientist frequently to estimate the total weight of an animal of given length, or meat weight of invertebrate for a given shell dimension. Consequently, determining the length–weight relationship is a standard procedure.

It is obvious that length and weight measurements used for a comparison must be obtained in the same way, but this is a problem of such complexity that it deserves a note here. The length of a fish is commonly measured in three ways: standard length; fork length, or length to the tip of the central rays of the caudal fin; and total length (Fig. 3.1). The first is used regularly for taxonomic purposes, the second is the common measure of the fishery scientist, the third is used occasionally. Morever, fish may be measured by the fishery scientist while alive, or immediately after death, or later during rigor mortis, or still later after rigor mortis. Fish are measured also commonly after preservation in formalin. It is important that measurements to be compared be the same measurements obtained under similar conditions or that they be adjusted. Live fish and fish in rigor tend to be shorter than relaxed fish after rigor. Preserved fish are also shorter because of shrinkage of the tissues. Weight measurements must be used also with caution because most animals lose fluids after death and add fluids after preservation in formalin or alcohol. Changes in both length and weight up to $\pm 5\%$, due to such factors, may occur.

With any set of lengths and weights the relationship is usually of the form

$$w = al^b$$

A line may be fitted easily to the data when they are transformed to (with logs to base 10)

$$\log w = \log a + b \log l.$$

This equation yields a straight line (Fig. 4.19).

When the exponent $b = 3$, the animal is growing without change in shape or specific gravity, i.e., isometrically. Among fish an exponent of 3 is the exception, however; values from 2 to 3.5 are commonly found.

Condition Factors.   A number of indexes of fatness, or condition, of aquatic animals have been developed. A direct index of fatness has been obtained by chemical analysis of fat content which is expressed as percent of body weight. A closely related index is the specific gravity of individual fish. This may be obtained if great care is used to eliminate all air bubbles. Those with lower specific gravity are fatter. Other direct indexes of fatness are based on the fat that surrounds the viscera. This may be judged subjectively or removed and weighed. A condition index of oysters is the ratio of the dry weight of the meat to the volume of the shell cavity $\times$ 100 (Fig. 4.18).

Other indexes are obtained by computation of the length–weight relationship. This may be shown as the estimated weight for a given length of fish (Table 4.2). Another index in common use for studies of freshwater fish is based on the cubic relationship

$$w = Kl^3,$$

in which K (usually multiplied by 100,000) is called the *condition factor*. The

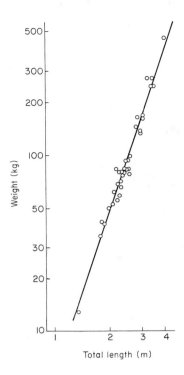

Fig. 4.19. Length-weight relationshp of Pacific blue marlin, *Makaira nigricans.* [Data from W. F. Royce (1957). Observations on the spearfishes of the Central Pacific. *U.S., Fish Wildl. Serv, Fish. Bull.* **57**(124), 495–554.]

TABLE 4.2

SEASONAL AND SEXUAL DIFFERENCES IN THE WEIGHT (IN POUNDS) OF MALE AND FEMALE YELLOWTAIL FLOUNDERS *(LIMANDA FERRUGINEA)* AT THE MEAN LENGTH OF 35.869 CM[a]

	Calender quarter	Male	Female	Difference[b]
	1	0.914	0.955	0.041
	2	0.826	0.945	0.119
	3	0.822	0.872	0.050
	4	0.882	0.953	0.071
Average	—	0.892	0.933	0.041

[a]Source: W. F. Royce, R. J. Buller, and E. D. Premetz (1959). Decline of the yellowtail flounder (*Limanda ferruginea*) off New England. *U.S. Fish Wildl. Serv., Fish. Bull.* **59**(146), 165–267.

[b]Note that females are always heavier and differ most greatly in weight from the males during the second quarter which is spawning season.

factor varies, of course, with the units of measurement; K is used commonly for grams and millimeters, C for inches and pounds. Either will vary also according to which length measurement is used.

4.7.3 ALLOMETRIC GROWTH

Even after the larval period, at the end of which the shape of most animals becomes similar to the adult, some changes in the body proportions continue to occur. Commonly they are minor changes, such as those in length of fins and fatness of body and the temporary changes associated with the ripening of the gonads. These changes are usually called *allometric growth*, although some grosser changes may be called *heterogonic growth*, such as the differentiation of right and left hand claws in crustaceans. The lack of change and the continuation of proportional growth is called *isometric* or *isogonic growth*.

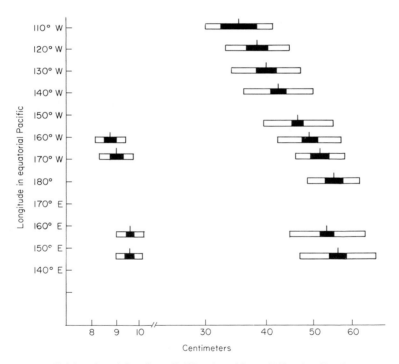

Fig. 4.20. Height of anal fin of small (65 cm) and large (140 cm) yellowfin tuna, *Thunnus albacares*, from the Pacific equatorial region as estimated from regression statistics. (The center line indicates the mean, the solid bar ± two standard errors of the mean, the hollow bar ± one standard deviation from regression.) [Source: W. F. Royce (1964). A morphometric study of yellowfin tuna *Thunnus albacares* (Bonnaterre). U.S., Fish Wildl. Serv., Fish. Bull. **63**(2), 395–443.]

Such changes in body proportions are of special interest to the fishery scientist who tries to differentiate stocks of the same or closely related species on the basis of body proportions. He must separate any differences due to growth from those due to either environment or inheritance. He can do so by comparing animals from different regions at identical lengths (Fig. 4.20). The yellowfin tuna, *Thunnus albacares*, of the example shows substantial variation in fin height due to locality in addition to great variations due to body size. Note that, if the anal fin grew isometrically after the tuna were 65 cm long, the fin would be only about 20 cm high when the tuna were 140 cm long instead of 30 to 65 cm.

Body proportions tend to vary more with growth than the numbers of body parts such as fin rays or vertebrae. The latter, called *meristic characters*, are more satisfactory for racial studies, but even they vary with growth occasionally.

4.7.4 RATE OF GROWTH

The essential way of expressing any kind of organic growth is to describe its rate during the life of the unit. This problem is one that has attracted the attention of many scientists who have sought general laws of growth by fitting mathematical functions empirically to growth data and by deducing other mathematical functions on theoretical grounds. As may be judged from the

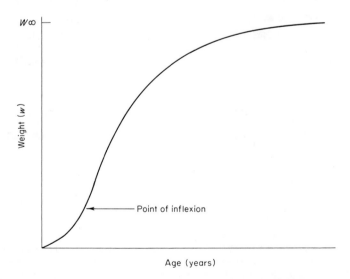

Fig. 4.21. Typical curve of growth in weight of aquatic animals. Point of inflexion is usually before half of the lifespan; thus an asymmetrical sigmoid curve results. In the von Bertalanffy growth curve (p. 241), the point of inflexion is at $w = 0.296\ W\infty$.

examples of growth given above and the discussion of the factors influencing growth, the rate may be the result of many factors operating independently and erratically. An adequate mathematical expression of the seasonal rate alone is extremely difficult, but even if the growth rate were assumed to be uniform throughout the year, the variations in yearly increments among species and populations would remain extremely complex. No generally applicable law of growth has been discovered yet.

One may say, however, that the growth of living aquatic units is almost always exponential; the rate is a function of the size already attained. Thus it is multiplicative or logarithmic rather than arithmetical. Early in the life of a unit, typically, the growth accelerates and later it slows. The curve of weight against age is at first concave upwards, then it flexes and becomes concave downwards as it gradually approaches an asymptote (Fig. 4.21). Even so, the rates are usually exponential; the early period positively, the later period negatively. Usually, too, the inflexion is early in life so the curve is an asymmetrical sigmoid with a long right-hand limb. (See Section 6.8.)

4.7.5 AGE DETERMINATION

It is essential to determine the age of animals for determination of the rate of growth as well as for the analysis of populations, which shall be discussed later. Fortunately, this can be done for most aquatic resource animals with relative ease. Three methods are available; the marked animal method and the length frequency method are used occasionally, and the annual ring method is used routinely. Each method has its special virtues and limitations so they frequently supplement one another.

The *marked animal method* requires that animals whose age is known initially be identified by a mark such as a tag or by isolation. When growth data are desired the size can be recorded at appropriate times, but the data must be interpreted with caution because the growth is almost certainly affected by the marking, handling, or isolation. The growth data may be less important, however, than evidence about the characteristics of the annual rings which can then be identified in other unmarked animals.

The second, or *length frequency method*, is used occasionally and is especially helpful for age determination of crustaceans and young animals, or determinations of seasonal growth. It is sometimes named for C. G. J. Petersen of Denmark, who introduced the method in 1891. This method is useful for broods of animals that have been spawned during a single, short period and that grow individually at nearly the same rate. When they are measured, the length frequency data form a mode or series of modes that can be identified. If, for example, collections were made of young fish, monthly, after their known birth date, it would be found that no other individuals of the same species would be near the same size, and the mode of this homologous group

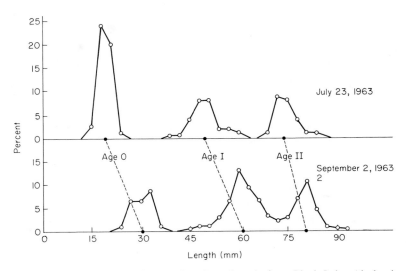

Fig. 4.22. Percentage length frequencies of pond smelt, from Black Lake, Alaska, in mid and late summer. (Unpublished data from Fisheries Research Institute, University of Washington, Seattle, Washington, 1900).

could be followed during successive collections (Fig. 4.22). With other data, a multimodal distribution may be found, in which two or three successive year classes can be identified.

The third, or *annual ring method*, which is by far the most useful, depends on the record left in the hard parts by irregular growth and metabolism. The irregularities are caused primarily by the seasonal changes in food, temperature, and spawning. Annual checks, or *annuli*, as they are commonly called, are usually the most prominent, but checks due to other causes, such as injury, can be identified sometimes. Furthermore, when the hard part grows at a known rate relative to the size of the body, the body size can be estimated for the time that each check occurred throughout life.

Many hard parts show checks, and the part to use may be chosen according to which is easiest to collect or examine, which has the clearest marks, which has the most consistent size relative to body size, or which may be collected without harm to the animal. The parts in most common use are the scales of fish (Fig. 4.23). Also useful from fish are the otoliths (ear bones) (Fig. 4.24), vertebrae, opercular bones, and fin rays. Similarly useful are shells of mollusks (Fig. 4.25), teeth of seals (Fig. 4.26), whalebone, and ear plugs of whales. Unfortunately, crustaceans have no permanent hard parts.

When the part is a fish scale, it may be examined directly or a plastic impression may be made of the surface. Either the scale itself or its impression may be examined under a microscope, or preferably, its image projected on a

Fig. 4.23. A scale of a mature sockeye salmon, *Oncorhynchus nerka*, from Bristol Bay, Alaska. Shown are the focus (F) of the scale, one annulus near the end of freshwater life (1) and three annuli during saltwater life (2, 3, 4). (Photo by Ted S. Y. Koo, Fisheries Research Institute, University of Washington, Seattle, Washington.)

Fig. 4.24. An otolith of a 12-year-old Pacific halibut, *Hippoglossus stenolepis*. (Photo by International Pacific Halibut Commission, Seattle, Washington.)

Fig. 4.25.   A numbered razor clam, *Siliqua patula*, added two clear annuli to its shell before recapture. (Photo by Herb C. Tegelberg, Washington Dept. of Fisheries, Olympia, Washington.)

screen (Fig. 4.27). When the part is a bone, tooth, or fin ray the checks may appear inside and it may be necessary to saw it into a thin section and polish this section for examination. Sometimes thin bones can be made translucent by chemical treatment and examined directly.

After any necessary preparation, the growth record in the hard parts is read. It is necessary to classify the kinds of checks and relate them to the actual events causing them. Annuli can be identified by a study of homol-

Fig. 4.26.   A longitudinal section of a canine tooth of a 9-year-old male fur seal (*Callorhinus ursinus*) taken at sea near Yakutat, Alaska, on April 21, 1959. Scale marks in millimeters. (Photo by David F. Riley, U.S. National Marine Fisheries Service, Seattle, Washington.)

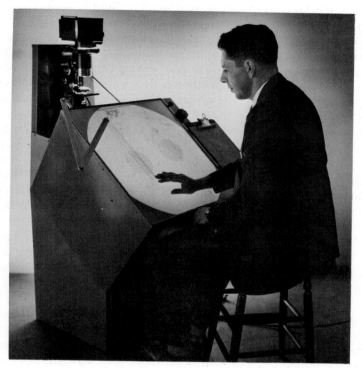

Fig. 4.27. A special microprojector is useful for reading and measuring fish scales. (Photo courtesy of Eberbach Corp., Ann Arbor, Michigan.)

ogous checks at the edge of hard parts collected at varying times during the year. Such a study permits a determination of time of completion of the annulus and also assists in identification of any checks due to other causes. As an alternative, parts may be studied from animals whose age has been determined by other methods. Special attention must be given to young animals since they may not form an annulus during their first year and to old animals because, frequently, they show poorly defined annuli. The process of reading is necessarily somewhat subjective, but it is fairly reliable when the reader is familiar with the life history of the group of animals with which he is working.

Age is usually designated as the number of annuli either in Arabic or Roman numerals. Some confusion exists because animals in their first year are designated "0" age before the first annulus is formed (which is usually close to 1 year after spawning). After the first annulus the count corresponds to human birthdays, i.e., an animal designated as age *3* is in its fourth year. A variety of special conventions is used when additional information can be obtained. Sockeye salmon, whose life in fresh water and in salt water can be recognized separately on their scales, are designated, for example, as age *2.2* when they have two freshwater and two saltwater annuli. This species of salmon fails to form an annulus in its first year to life so this example would be 5 years old. Another common method of designating the age of this example is 5_3; 5 shows total age and the subscript is the number of total years in fresh water.

Much more information than age at capture can be obtained from the hard parts when, as was mentioned earlier, the relationship of hard part size to body size is known. The size at all earlier "birthdays" can be estimated. Usually the easiest points to measure are along a radius to the boundaries marking the completion of each annulus and the beginning of the following year's growth (Fig. 4.28). Such measurements frequently provide extremely valuable information on the early life of the animals.

Unfortunately, the growth of hard parts is rarely isometric. Scales and many bones do not form in the larva until the body reaches an appreciable size and may grow subsequently at different rates. Some early investigators assumed isometric growth and soon discovered puzzling discrepancies between the size estimated from hard parts and the size of young animals actually caught. The most common result is the tendency for the estimated length of a young animal of specific age to decrease as the total age at capture increases. This has been called *Rosa Lee's phenomenon*, after its discoverer. It is caused by allometric growth of hard parts (Fig. 4.29) and by more subtle factors, including a marked tendency of fishing gear to select fish of certain

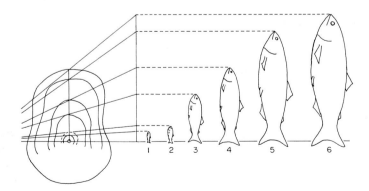

Fig. 4.28. Diagram of approximate relationship of salmon size to scale radii. The length at capture (6) and the scale radii can be used to estimate length at the times of annulus formation (1, 3, 4, 5) and the time of migration to sea (2).

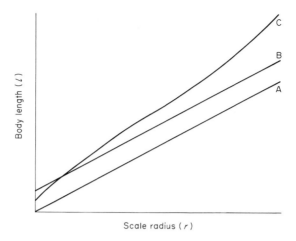

Fig. 4.29. Types of relationship between fish scale and body length. (A) Isometric growth, $l = br$. (B) Allometric growth typical of late-appearing scale, $l = a + br$. (C) Allometric growth of type that must be fitted with curvilinear relationship.

size. Commonly, larger young fish and smaller old fish are taken by a single kind of fishing gear—a circumstance that alone can cause Rosa Lee's phenomenon.

4.8 Toxins and Diseases from Aquatic Resources

Unfortunately aquatic organisms pose many hazards for humans who use the water or its resources for themselves or their domesticated animals. Many vicious, poisonous, and venomous organisms may be encountered in the water, along with many human diseases that are waterborne.

Venomous animals possess special toxin-producing glands associated with an apparatus for delivering the toxin. Such animals are found among hydroids, jellyfishes, sea anemones, starfish, sea urchins, cone shells, octopi, sting rays, and fishes, principally of the catfish and scorpion fish groups. Poisonous organisms contain materials in their body that will poison an animal consuming them.

Another kind of hazard arises from fish and shellfish that carry disease producing organisms acquired either through the water in which they live or by contact after death. Mollusks that are consumed raw may transmit human diseases if they have been taken from polluted waters. Mollusks, crustaceans, and fish may all be contaminated after death if they are not thoroughly chilled or if they are handled under unsanitary conditions. Still other hazards arise because of allergic type reactions. Sea foods are available in such variety that many people are likely to be allergic to some kinds. A discussion of either the problem of disease transmission or the allergenic reactions is beyond the scope of this text.

Any of the toxic or venomous organisms may become resource organisms if a use is discovered for their toxic materials. Our constant search for useful drugs commonly starts with substances known to have a powerful effect on animals. Such substances frequently have beneficial effects at sublethal dosages. The current attempts to identify drugs from the sea include an examination of many toxins and venoms, some of which have pharmacological properties like anesthetics.

4.8.1 FISH POISONING

Fish which are commonly or always toxic are usually avoided by knowledgeable persons. They may, of course, poison the uninformed and experimenting person, but they are rarely responsible for widespread trouble.

Much more serious are the species of fish that are regularly eaten and which may be toxic only at certain places or at certain times. These have been responsible for many cases of poisoning and hundreds of fatalities. Several hundred marine species, most of which live in warm, shallow waters between 35° North and 35° South latitudes, have been implicated. Included are some choice food fishes of the families Carangidae, Lutjanidae, and Serranidae. Some freshwater species are known to have toxic roe but not toxic flesh.

Much of the fish poisoning is called ciguatera. The symptoms, usually evident within 4 hours, are nausea, numbness about the mouth and throat, weakness, abdominal pain, and chills. Fatal cases also involve a severe fall in blood pressure and respiratory failure.

The cause of the toxin in the fish is unknown although it is suspected of originating in toxic bottom-growing algae. Many of the toxic species of fish are either feeders on bottom algae or carnivorous on other fish which presumably fed on the toxic algae. The most poisonous parts of the fish are the liver and the gonads, and these parts may be poisonous when the flesh is not.

Another kind of fish poisoning occurs following the consumption of puffer fish of the family Tetraodontidae. A large proportion of such fish are poisonous. The clinical symptoms are similar to ciguatera, but the poison appears to be more potent. Several species of the puffers are consumed in Japan where they are called fugu and served in especially licensed places to people who desire the anesthetic effect. The practice is dangerous and, despite the attempts at control, it results in a few fatalities each year.

Unfortunately the toxins involved in fish poisoning cannot be detected by any simple chemical test. The best test is an intraperitoneal injection of an aqueous extract in laboratory mice that are observed subsequently for their reaction. Nor can the toxins be destroyed by any known treatment without also destroying the flesh. As a consequence, some species of excellent tropical food fish cannot be marketed on a large scale because of the possibility of an occasional toxic one.

4.8.2 PARALYTIC SHELLFISH POISONING

About 30 species of mollusks including oysters, clams, and mussels are known to have caused paralytic shellfish poisoning. The symptoms are similar to those of ciguatera, but usually occur within a half hour after eating the mollusk. Fatalities occur occasionally.

Most of the studies of paralytic shellfish poisoning have shown that it is related to the occurrence of dinoflagellates of the genus *Gonyaulaux*. These are consumed by the mollusks which accumulate and store the poisons for varying lengths of time. Usually the dinoflagellates bloom during the summer months and the mollusks are toxic for a short time following the bloom. However, the butter clam, *Saxidomous*, found on the west coast of North America, stores the poison for a long time and may be toxic the year round.

The cases of paralytic shellfish poisoning have occurred mostly in the subarctic to temperate waters of the northern hemisphere on the west coast of North America and off Northwest Europe. Other locations include waters off Eastern Canada, Japan, New Guinea, and New Zealand.

Where beds of edible mollusks are suspected of being poisonous, they are tested carefully by public health officials and the area closed to harvest during periods when they are toxic (usually during the summer). Such beds occur along the outer coasts or in the broad straits along the west coast of North America.

4.8.3 FISH AS INTERMEDIATE HOSTS OF DISEASE ORGANISMS

In addition to serving as carriers of disease organism acquired from the water or from contamination due to unsanitary handling, a few aquatic resource animals may act as an intermediate host of a disease organism that can be transmitted to humans.

One disease afflicts people in northern North America, Europe, and Asia who eat uncooked freshwater fish. This is a cestode, the broad fish tapeworm, *Dibothriocephalus latus*, that reaches an adult stage between 3 and 10 m long in the intestine of fish-eating mammals, including man. The adults discharge eggs in huge numbers which hatch in the water, they are eaten by certain copepods which in turn are eaten by fish, and they in turn are eaten by other fish. The larval stages of the worm live in the musculature of the fish and reach the mammal when the fish are eaten.

Other widespread diseases of humans occur in Eastern and Southeastern Asia. These are caused by trematodes or flukes of several species. One lives as an adult in the bile ducts of cats, dogs, or man and infests a snail and a cyprinid fish during its larval stages. Others live as adults in the intestine of fish-eating birds and mammals, including man, and infest snails and mollusks or cyprinid fishes during larval stages. Another is a lung parasite of man that infests a snail and freshwater crustacean during its larval stages.

Both the cestode and the trematode parasites require freshwater animals as intermediate hosts and are transmitted to man if he eats uncooked or partially cooked fish, mollusks, or crabs from fresh water. By contrast, no saltwater fish, mollusk, or crab is known to be an intermediate host for a significant human disease, and may be eaten raw without fear of worm infection.

REFERENCES

Borradaile, L. A., Potts, F. A., Eastham, L. E. S., and Saunders, J. T. (1963). "The Invertebrata," 4th ed. Cambridge Univ. Press, London and New York.
Brown, M. E., ed. (1957). "The Physiology of Fishes," Vols. 1 and 2. Academic Press, New York.
Cahn, P. H., ed. (1969). "Lateral Line Detectors." Indiana Univ. Press, Bloomington.
Carlander, K. D. (1969). "Handbook of Freshwater Fishery Biology," Vol. 1. Iowa State Univ. Press, Ames.
Halstead, B. W. (1965). "Poisonous and Venomous Marine Animals of the World," Vols. 1 and 2. US Govt. Printing Office, Washington, D.C.
Hasler, A. D. (1966). "Underwater Guideposts, Homing of Salmon." Univ. Wisconsin Press, Madison, Wisconsin.
Hoar, W. S., and Randall, D. J., eds. (1969, 1970). "Fish Physiology," Vol. 1–4. Academic Press, New York. Other volumes scheduled.
Kinne, O. (1963). The effects of temperature and salinity on marine and brackish water animals. I. Temperature. *In* "Oceanography and Marine Biology" (H. Barnes, ed.), Vol. 1, pp. 301–340. Allen & Unwin, London.
Kinne, O. (1964). II. Salinity and temperature salinity combinations. *In* "Oceanography and Marine Biology" (H. Barnes, ed.), Vol. 2, pp. 281–339. Allen & Unwin, London.
Kleerkoper, H. (1969). "Olfaction in Fishes." Indiana Univ. Press, Bloomington.
Lagler, K. F. (1956). "Freshwater Fishery Biology," 2nd ed. W. C. Brown, Dubuque, Iowa.
Mistakidis, M. N., ed. (1967). "Proceedings of the World Scientific Conference on the Biology and Culture of Shrimps and Prawns," Vols. I, II, and III. FAO, Rome, Italy.
Moulton, J. M. (1964). Underwater sound: Biological aspects. *In* "Oceanography and Marine Biology" (H. Barnes, ed.), Vol. 2, pp. 425–454. Allen & Unwin, London.

Needham, A. E. (1964). "The Growth Process in Animals." Van Nostrand-Rheinhold, Princeton, New Jersey.

Nicol, J. A. C. (1967). "The Biology of Marine Animals," 2nd ed. Wiley, New York.

Nikolsky, G. V. (1963). "Ecology of Fishes." Academic Press, New York. (Translated from the Russian.)

Norman, J. R. (1963). "A History of Fishes" (2nd edition by P. H. Greenwood). Ernest Benn, London.

Paloheimo, E. H., and Dickie, L. M. (1965). Food and growth of fishes. Vol. I. A growth curve derived from experimental data. Vol. II. Effects of food and temperature on the relation between metabolism and body weight. Vol. III. Relations among food, body size, and growth efficiency. *J. Fish. Res. Bd. Can.* **22**(2), 521–542; **23**(6), 869–908 (1966); **23**(8), 1209–1248 (1966).

Purchon, R. D. (1968). "The Biology of the Molluscs." Pergamon, Oxford.

Rosa, H., Jr., ed. (1959). "Proceedings of the World Scientific Meetings on the Biology of Sardines and Related Species," Vols. I, II, and III. FAO, Rome, Italy.

Rosa, H., Jr., ed. (1963). "Proceedings of the World Scientific Meetings on the Biology of Tunas and Related Species," Fish Rep. No. 6, Vol. I–IV. FAO, Rome, Italy.

Sette, O. E. (1943). Biology of the Atlantic mackerel (*Scomber scombrus*) of North America. Part I. Early life history, including the growth, drift, and mortality of the egg and larval populations. *U.S., Fish Wildl. Serv., Fish. Bull.* **50**, 149–237.

Waterman, T. H. ed. (1960). "The Physiology of Crustacea," Vols. 1 and 2. Academic Press, New York.

Woodhead, P. M. J. (1966). The behavior of fish in relation to light in the sea. *In* "Oceanography and Marine Biology." (H. Barnes, ed.), Vol. 4, pp. 337–403. Allen & Unwin, London.

Basic Ecology

The fishery scientist is concerned with more than individual living organisms. He is concerned with groups of individuals of the same species called *populations*,* with groups of populations of different species in given areas called *communities*, and with the functioning together of communities with their nonliving environments as *ecosystems*. An individual, a population, a community, or an ecosystem can be considered as a biological unit.

The scientist is confronted with ecosystems whether he is trying to maximize production from wild populations, husband an artificially confined population, or predict the effects of a changing environment on a community. He is faced, in a general sense, with problems in aquatic ecology.

Ecology is a broad, complex, and very young science. It is a fundamental branch of biology that is defined as the interrelationships among organisms and the interrelationships of organisms with their nonliving environment. It embraces the terrestrial environment as well as the aquatic. That part of ecology that deals with the interrelationship involving insects, higher plants, birds, and terrestrial mammals will not be reviewed in this chapter, but the principal interrelationships involved in producing food or recreation from the aquatic environment will be summarized.

* Population is used in an ecological sense to mean a defined group of the same species larger than a family group. It is also used in a geographic sense to mean a group occupying a defined area or, more generally, to mean a group defined on the basis of any criterion.

5.1 Single-Species Populations

The reader may have noted earlier in this book that the factors of the environment have been discussed with respect to their effect on living organisms and the animals have been discussed with respect to their reaction to factors in the environment. All of this discussion has been an introduction to the central fishery problems of producing food and recreation from the living aquatic resources as parts of ecological systems. The next step toward considering the complexities of the systems is to discuss the characteristics of single-species populations. Such a population never exists in nature because no organism exists alone; an organism, even when it is apparently isolated, may have parasites, and every organism is dependent upon another organism, either living or dead, for food. Nevertheless, much of the theory about ecology has developed from studies of single-species populations, and in fishery practice, many of the concepts that have been developed from single-species populations are applied directly and usefully to the study of stocks being exploited, under the assumption that each stock lives in a constant environment.

5.1.1 LIMITING FACTORS

Every population of a species of organisms is limited in its distribution and abundance by its relative ability either to utilize or to tolerate the factors in its environment and by its relative ability to disperse. Each population lives in a *habitat* and it also has a *role* among the other populations in its community. Its habitat and role together form its *niche*. The niche includes the relations of the population to food, to predators, to competitors, and to physical environment at all stages in its life. The niche of young individuals may be entirely different than that of old and may vary from one environment to another. For example, the niche of sockeye salmon includes its role in a gravel stream bottom for about 7 months, a lake for 1 or 2 years, coastal ocean waters for about 3 months, the surface layer of the high seas for 2 or 3 years, and a stream for the final month of its life.

As organisms have evolved to tolerate the multitude of environments on earth, each species has found a niche of its own, not completely occupied by any other species although partially occupied by a multitude of other species. Neither in nature nor under experimental conditions do two species occupy exactly the same niche for long. Even when two similar species live together in the same place they will differ in some aspect such as food, tolerance to some environmental factor, or breeding requirements, and one will tend to replace the other. This singularity of one species in one niche is known as the *Volterra-Gause* or *exclusion principle*.

Each species has a set of ideal conditions under which it thrives, but it can tolerate more or less change from the ideal conditions. The degree of tolerance to a factor is expressed by the prefixes *eury-*, meaning more tolerant, and *steno-*, less tolerant. For example, a eurythermal animal tolerates a relatively large range of temperature, a stenohaline animal a relatively small range of salinity.

Each species has its own set of limiting factors. An aquatic species is limited by temperature, salinity, depth, light, dissolved oxygen, and probably by any of many other factors. Many of the elements present in minute quantities, poisons excreted by certain algae, interactions with virus or bacteria have been shown to be limiting. Anything in the environment that affects the physiology of an organism may be at some level a limiting factor for that organism.

The limiting factors operate according to their strength or level in complex ways. Most physical and chemical factors have maximum and minimum limits. Some factors may operate only at certain times, others continuously. Usually, one factor will be more critical than others and will be the controlling factor. Among required chemical factors the controlling factor will be the one available in the smallest fraction of the amount needed. This is called *Liebig's law of the minimum* and has been extended in an abstract sense to physical and biological factors.

The reproductive period is commonly the time of least tolerance. The conditions must be tolerable by spawning animals, eggs, larvae, and young which frequently have different needs. For example, the Pacific oyster requires a minimum temperature for successful spawning but will grow and fatten perfectly well at much lower temperatures. This limiting factor is overcome in oyster culture by collection of the young in spawning areas and their transfer to growing areas.

The determination of tolerance levels is important in understanding the distribution of organisms and is receiving increasing attention as man changes the environment. With many chemical and physical factors, tolerance is a function of both level and time; the greater the change from the ideal level, the less the length of time the change will be tolerated. Potential limiting factors are tested at gradients of concentrations for determination of the time required for manifestation of either discomfort, change in behavior, change in body chemistry, or, more commonly, death of a fraction of the organisms. The easiest criterion to use is the time of death for half of the animals, but this level is obviously different from the tolerable level; nevertheless it is a useful index.

Even though one factor may be critical in a certain situation, often, factors operate together to create *synergistic* effects. For example, an animal contending with low oxygen is almost certainly less able to tolerate either

TABLE 5.1
LETHAL SALINITIES FOR LOBSTERS SUBJECTED TO VARIOUS COMBINATIONS OF
TEMPERATURE AND OXYGEN[a]

Oxygen (mg/liter)	Salinity (%₀)						
	5°C	9°C	13°C	17°C	21°C	25°C	29°C
1.0	18.0	21.0	20.4	22.0	30.0	—	—
2.0	13.3	12.6	12.0	11.2	12.0	15.0	—
3.0	11.6	10.4	9.6	9.0	9.3	11.2	30.0
4.0	11.4	9.7	9.4	9.0	9.3	11.2	20.4
5.0	11.0	9.5	8.8	8.8	9.3	11.2	17.8
6.0	9.8	8.8	8.4	8.4	9.3	11.2	16.4

[a]When the lobsters were subjected *gradually* to each of the combinations indicated, their average mortality rate was 50 percent in 48 hr. [Source: D. W. McLeese (1956). Effects of temperature, salinity, and oxygen on the survival of the American lobster. *J. Fish. Res. Bd. Can.* **13**, 247–272.]

high temperature, or low salinity, or shortage of food. An excellent example of the operation of three factors on lobsters is given by McLeese (1956) in Table 5.1. Lobsters that prefer cool, well-oxygenated seawater tolerate the highest temperature under conditions of high oxygen and high salinity, the least oxygen under conditions of high salinity and low temperature, and the lowest salinity at medium temperature and high oxygen.

Testing the tolerance for chemical and physical factors is rendered difficult, not only by the synergistic effects, but also by the ability of animals

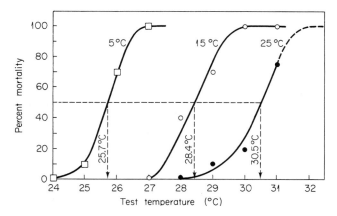

Fig. 5.1. Mortality rates at 48 hours for lobsters acclimated at 5°, 15°, and 25°C each at 30%₀ salinity and 6.4 mg O₂/liter. Fifty percent lethal temperatures are shown on each curve. [Source: D. W. McLeese (1956). Effect of temperature, salinity, and oxygen on the survival of the American lobster. *J. Fish. Res. Bd. Can.* **13**, 247–272.]

to become *acclimatized* and to be more tolerant afterward (Fig. 5.1). (The lobsters tested for Table 5.1 were fully acclimatized, or else they would have been far less tolerant.) The mechanisms for acclimatization are little understood but are common to many animals. Aquatic animals may acclimate to higher pressure, extreme temperature, low oxygen, and even to certain poisons.

Another kind of long-term adjustment occurs in populations where those individuals that are better able than others to tolerate certain factors survive and transmit their characteristics to their offspring—the process of natural selection. Such *adaptations* occur increasingly as man changes the environment and creates new niches. Adaptations that have been demonstrated include resistance to disease, improved ability of predators to find and consume food, improved ability of prey to avoid and resist predators, tolerance of chemicals, and avoidance of fishing gear. Some species adapted extensively to chemical and biological stress in five to eight generations in experiments.

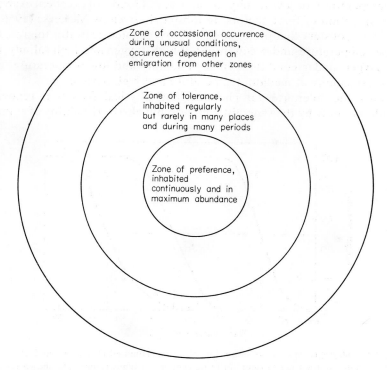

Fig. 5.2. Diagram of distribution of organisms. [After Andrewartha and Birch (1954). Copyright 1954 by the University of Chicago Press.]

After each population of each species of an organism has acclimated to the maximum amount and has adapted as a result of natural selection, it is still limited in distribution and abundance by some levels of critical factors. Its distribution will tend to be like that diagrammed in Fig. 5.2, with a region where abundance is maximal, a region where occurrence is regular but not abundant at all times, and a region where appearance is occasional and temporary. In the aquatic environment such a distribution is always three-dimensional. The boundaries are complex and are usually related to different factors in different places. For example, the distribution of a coastal marine bottom fish might be limited by deep water on the ocean side, high temperature at the south end, wave motion on the land side, absence of food above the bottom, low salinity in an estuary, and a current that carries its eggs out to sea on the north side.

5.1.2 PATTERNS OF DISTRIBUTION

Any population has a characteristic structure or arrangement of its individuals, known as its *pattern of distribution*. The pattern of distribution results from the entire behavioral response of individuals to the factors of the environment. Patterns vary greatly in scale from a few times the length of the individual to thousands of kilometers. They are not as orderly as the distribution of cells in an organism but are orderly enough to be classifiable.*

One of the kinds of patterns is called *vectorial*; it has already been discussed at some length but not named. This is the distribution of individuals in response to the chemical and physical factors of the environment such as temperature, light, pressure, salinity, current, and bottom type. With most such factors, individuals occur in numbers according to the level of the factor. For example, the vectorial distribution in Fig. 4.3 is in response to light.

A second kind of pattern, the *reproductive*, is evidenced for a time by the species that care for their young. Some species of fish any many of the aquatic mammals form family groups for a few days and some for many months. Another kind of distribution associated with reproduction is the pattern of distribution of the young after dispersal by the currents. Many marine species of crustaceans, mollusks, and fish have such pelagic young, and their distribution is the result of spawning time, spawning place, drifting time, and the set of the ocean currents.

Aggregations of a species for spawning, feeding, or other *social* purposes is another kind of pattern of special importance to the fisheries. This pattern has already been discussed under the topic of schooling (see Section 4.4).

---

*This classification follows that by G. E. Hutchinson (1953). The concept of pattern in ecology. *Proc. Acad. Natur. Sci. Philadelphia*, **105**, 1–12.

TABLE 5.2
HYPOTHETICAL SERIES OF CATCHES FROM OVERDISPERSED,
RANDOM, AND CONTAGIOUS DISTRIBUTIONS

Sample number	Overdispersed (nearly uniform)	Random	Contagious
1	5	4	0
2	5	7	0
3	4	5	0
4	5	6	45
5	6	8	0
6	5	2	0
7	5	5	5
8	4	6	0
9	6	3	0
10	5	4	0
ΣX	50	50	50
$\overline{X}$	5	5	5

The small-scale dispersion of a species in a uniform part of its environment will tend to be a *random* or a stochastic pattern. A random distribution is a result of chance, but frequently, social factors result in a distribution of clumps of animals that may be random in size and form random distributions themselves. A distribution of clumps of individuals is called a *contagious* distribution, or *underdispersion*. The opposite distribution, *overdispersion*, (Table 5.2) is rarer than the random but might occur in samples from a school of uniformly spaced animals. (The reader should note that some ecologists have used superdispersion to mean underdispersion and infradispersion to mean overdispersion.)

The description of an animal distribution usually depends on the fact that a series of random samples of randomly distributed animals will contain a frequency distribution of animals per sample that conforms mathematically to the *Poisson* distribution. This distribution is the expansion in which

$$\frac{1}{e^m}, \frac{m}{e^m}, \frac{m^2}{2!e^m}, \frac{m^3}{3!e^m}, \frac{m^4}{4!e^m} \cdots$$

e is the base of natural logarithms and m is the mean frequency of occurrence. When a comparison of a frequency distribution with the expected Poisson leads to the conclusion that contagion exists, other mathematical distributions, such as the negative binomial, may be appropriate. [For a further discussion and extensive bibliography, see Southwood (1966).]

The fifth and last type of pattern to be mentioned here is the *coactive*; it results from competition between closely related species. It follows from

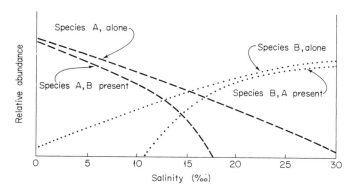

Fig. 5.3. Diagram of the vectorial distribution of each of two species alone in a salinity gradient and their coactive distribution when existing together.

the exclusion principle (see page 165) that only one such species can occupy a single uniform environment, but it happens in nature that few environments are uniform. Almost always, two or more competing species can exist together because each can find its niche in a part of the environment that it can use better than the others. The two or more species will tend to reach an equilibrium distribution that depends on the species involved and on the gradients in the physical features of the environment.

The resulting distribution is in part vectorial and in part coactive. This distribution is typical of the distributions of many organisms. To illustrate, let us assume that a single species, A, exists in a salinity gradient that restricts but does not prohibit the occurrence of the species at the high salinity end; and we find a vectorial distribution of species A as shown in Fig. 5.3. If we added a competing species, B, that favors high salinity, we would expect that it will predominate at the high salinity, we would expect that it will predominate at the high salinity end and depress the abundance of the species that favors low salinity.

5.1.3 DYNAMICS

The study of a population as a living unit is usually called *population dynamics* because interest is focused on changes in the population associated with birth and death rather than on its static composition. There are, however, numerous similarities and contrasts with a living individual (Table 5.3). A population is a unit in a space with a size, usually expressed as a density or abundance. It maintains itself in continuous life as a rule; birth and death of populations occur only rarely. It has, however, a natality (birth rate) and

TABLE 5.3
COMPARATIVE CHARACTERISTICS OF INDIVIDUALS AND
POPULATIONS

Individual	Population
Size	Density, abundance
Morphology	Pattern of distribution, age structure
Growth in length or weight	Growth in number or biomass
Birth	Rate of natality
Death	Rate of mortality

a mortality (death rate). It has a physical structure, expressed as its pattern of distribution (discussed in Section 5.1.2), and it has an age structure. It grows at times in ways remarkably like those of growth of an individual, but it may decrease in size also. Most important to the fishery scientist, it has a capacity for production.

Density. The size of a population is expressed by the total number or total weight. The latter, commonly called the *biomass*, may be expressed as either the live weight or the dry weight. Neither number nor weight is usually measurable directly, therefore relative number and weight or indexes are used; these are called the *density*, or *abundance*. The common unit in fisheries is the *catch per unit of effort* by a specified kind of fishing or sampling gear. When the gear either covers a known area or strains a known volume of water, the catch per unit of effort may be transformed to a catch per unit of area or volume.

Such relative measures of density are useful when indexes of the same type are being compared. One of the central problems of population dynamics is converting the relative measure into absolute numbers or weight.

Natality is the rate at which new individuals are added to a population by reproduction. It is a broader term than birth rate, because it includes young that result from eggs, spores, or seeds, as well as those that are born. It is less than the rate of fecundity of egg-, spore- or seed-producing organisms because individuals are not usually counted until after hatching or germination. Natality of aquatic resource populations is rarely measurable, however, because the new individuals are so tiny, so dispersed, or dying so rapidly. Instead, the addition of young animals of a size catchable by fishing gear is measured and called the *recruitment*.

Mortality. The number of individuals in a population decreases by death, and because every organism usually produces a surplus of young, the rate of mortality at early life stages is usually the primary factor that regulates the number in a population.

The direct causes of natural death in aquatic populations are varied and not well known. Dead aquatic animals are seen rarely and dying animals more rarely. Even in fish catches from populations with known natural mortality rates ranging from 20 to 30%/year, the proportion of dying to healthy individuals seems much less than the expected 1 or 2/1000/day.

In fresh waters of temperate zones dead fish are observed most commonly in spring because at this time fish seem less resistant to disease and dead bodies decompose less rapidly than at other seasons of the year. In summer and fall fish still die, but those that die are consumed rapidly by scavengers. Small animals are commonly eaten by predators, but larger animals more rarely. Epidemics of either disease or parasites are common, even a constand menace, among cultivated populations of fish or mollusks, but appear to be rare in natural populations. Occasionally, windrows of stinking animals may be visible on the beaches from either natural catastrophes, such as shifts in ocean currents, oxygen depletion under ice, or biological catastrophes such as blooms of poisonous plankton (red tide); but generally, mortalities among aquatic animals, even in large populations, are unnoticed and uncounted.

A consequence of our inability to observe death directly is our inability to estimate the rate of mortality directly. Almost always it is determined from the number of survivors at various times from a group of individuals that started life at the same time. Such an estimate is attended usually by considerable difficulty (see Section 6.7) and can be made rarely at close intervals, so in practice, an estimate of mortality rate spans a considerable period of time. Despite the certainty of death its measurement is somewhat uncertain.

Regardless of circumstances, no population increases indefinitely; each is limited in size by some mechanism that either decreases the rate of natality or increases the chance of death for the individual as the population increases. Such a mortality is said to be *density-dependent* or *compensatory*. A number of somewhat interrelated mechanisms that regulate the mortality in this way have been identified.

Intraspecific competition for the essentials of life is the first and probably the most important limiting mechanism. As organisms increase in number, their food, space to move in, opportunities to get rid of waste, ability to avoid predators, and opportunities to transmit disease change in ways that tend to increase mortality rate.

A second mechanism is an increase in environmental heterogeneity. As a population increases in number, it tends to occupy a larger area, and this area usually includes zones that are less than favorable for the population

(Fig. 5.2). Such zones may be uninhabitable during times of unusual conditions, and the organisms in them must migrate or die.

A third mechanism is an increase in predator (including parasite) populations as a consequence of a greater number of prey and a greater ease of finding them. Such a buildup of a predator population tends to lag behind the increase in the prey population so that when the predation becomes limiting to the prey population, the predator population is still increasing. The consequence is likely to be a major decline in the prey population.

Another mechanism, the genetic adaptation of populations, operates to enhance the survival of each succeeding generation under stress. When the population is a prey species, it tends to become more adept at evading the predator. When it is either a predator or a parasite species, it tends to become more adept at catching the prey. When it is subject to a limiting factor in the environment, it tends to overcome it to some extent. The result in a varying environment is a dominance, at least temporarily, of the more quickly adaptable populations.

The effects of all of these mechanisms change regularly with season, irregularly with other climatic and biological factors, and according to the age of individuals. Predators are more active at certain times than at others and shift their preference from one food to another frequently. Usually food supplies vary during the seasons and probably affect directly the well being of the feeders. Among the stages of animals, egg and larvae are probably most in danger (note that $10\%/$day, or higher, mortality rate of young mackerel in 1 year, see Section 4.6.9); young adults are probably subject to the least chance of death, and older adults subject to a greater chance of death. The consequence of all these variable factors is a large variability in number in the population.

5.1.4 AGE STRUCTURE

The interplay of natality and mortality in a population results in a set of proportions of the different age groups. Age may be measured in any unit of time, but for fishery resource populations, a year is almost always the appropriate unit, especially when the individuals are born during a single season of the year.

Much of the usefulness of the age structure depends on its relationship to the survival rate of a *cohort* or *age group* (a group of animals of the same age). In a stable population the proportion of each age group at a given time is equal to the proportion of each corresponding age during a cohort's life. Because it is far quicker to obtain estimates of the proportion of each age group in a population of a given time than to wait out a cohort's life, this relationship is frequently used. From it are constructed life tables of the

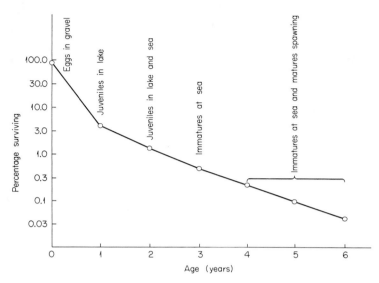

Fig. 5.4. Estimated approximate age structure of sockeye salmon from Bristol Bay on September of each year after spawning and before death of mature individuals.

estimated survivorship of cohorts, and from these may be computed estimates at each age of numbers dying and mortality rates and life expectancies.

Despite the ease of age determination of most fishery resource animals, complete age structure has been determined for few populations. The reasons lie in the separate distribution of eggs, larvae, juveniles, and adults for most animals and in the selectivity of all fishing or sampling gear for animals of certain sizes. For salmon, survival rates during the life in the gravel, the life in the lakes, and the life in the sea have been estimated with reasonable accuracy, but even in its case it is necessary to assemble an approximate composite (Fig. 5.4). As the figure shows, in salmon the survival rate during the egg and larval stages is lower than during the other stages; among animals having larger numbers of eggs it is lower still. The survival rate is nearly uniform among immature and adult salmon; among many species of fish this trend in survival rate is typical.

Although complete age structure is nearly impossible to determine, *age compositions* of the catches of sampling or fishing gear are easy to determine and are of much use in the analysis of fishery resources. An age composition of halibut, a long-lived species, from a catch in a virtually unexploited population is an example (Fig. 5.5). The figure shows the typical feature of absence or a lower proportion of the younger age groups than of the older;

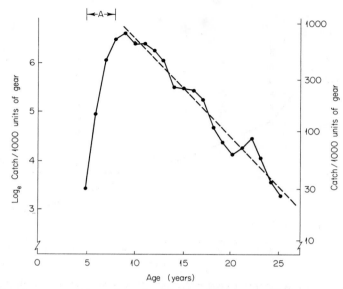

Fig. 5.5. Age composition of catch of halibut from an unfished area in the Bering Sea, June, 1930. (A) shows the ages of reduced availability to the long line gear. The dashed line indicates a constant instantaneous annual mortality rate of 0.235 (see page 236). [Data from: H. A. Dunlop, F. H. Bell, R. J. Myhre, W. H. Hardman, and G. M. Southward (1964). Investigation, utilization and regulation of the halibut in southeastern Bering Sea. *Int. Pac. Halibut Comm. Rep.* **35**, 1–72.]

the younger age groups are not fully available to the fishing gear either because they are distributed differently from the older animals or because they cannot be caught as efficiently as the older animals.

5.1.5 Population Growth

When a species is introduced into a new environment it grows in number and actual weight at a rate that resembles the sigmoid (S-shaped) growth of an individual organism (Fig. 4.21). At first the introduced organisms must acclimatize and complete their reproductive cycle. Then, if there are few environmental restraints, their natality will greatly exceed their mortality and the number in the population will increase exponentially for a time. Later, as food, space, disease, or some other factor restrains their numbers, the rate of increase will slow and the population will approach a size at which average natality and average mortality are equal. This population size will not be a limit in the mathematical sense, like the limit on the size of the individual, but a point of balance with more or less fluctuation around it.

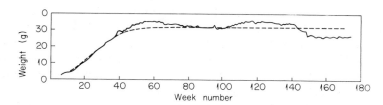

Fig. 5.6. Growth in biomass through 175 weeks of a population started from 5 pairs of guppies in an aquarium. The dotted line is a fitted logistic curve. [After Silliman and Gutsell (1958).]

Such a growth pattern has been demonstrated experimentally with many short-lived organisms. Experiments (Silliman and Gutsell, 1958) in which populations were started in aquaria with 5 pairs of guppies each and kept under constant conditions with a constant amount of food show clearly this pattern of growth (Fig. 5.6). For a few weeks, the total weight of guppies in an aquarium accelerated, then slowed, and finally reached equilibrium at about 30 g and subsequently fluctuated about ± 5 g. Since there was no other species of fish in the aquaria, this situation was an example of intraspecific competition.

Similar patterns of population growth occur in nature when animals are introduced into a new environment. Examples of such patterns among fish

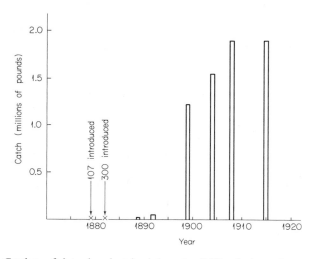

Fig. 5.7. Catches of introduced striped bass in California in various years between 1888 and 1915. After 1915 catches were lower, but fishery was restrained by law. [After J. A. Craig (1930). An analysis of the catch statistics of the striped bass (*Roccus lineatus*) fishery of California. *Div. Fish Game Calif., Fish. Bull.* **24**, 1–41.]

are the growths of populations from striped bass (Fig. 5.7) and the shad introduced into Pacific coastal waters from the Atlantic coast. Other examples are the growths of populations from pests that have been introduced accidentally. The American oyster drill, *Urosalpinx* (a snail called tingle in England), is the most serious predator on English oysters, and the American slipper limpet, *Crepidula*, is the most important competitor of oysters. The former was introduced about 1910 and the latter about 1880; both have caused much greater problems for oyster growers than similar native animals.

Exponential population growth in nature may occur seasonally. Examples are the blooms of plankton in the spring in temperate zones. A typical sequence of events is the warming and enrichment of the water, the accelerated population growth of a few species of phytoplankton, and the accelerated population growth of a few species of zooplankton. With both kinds of plankton the overwintering forms are released from their environmental restraints and stimulated to reproduce at near maximum rates until restraints become effective again.

5.2 Communities and Ecosystems

Any population of a species outside a laboratory or a farm lives as a part of a *community* of species on which it depends and which depends on it. It may be a producer (a plant population) that carries on photosynthesis, a *consumer* (an animal population) that feeds on plants or animals, or a *decomposer* (a bacterium or fungus population) that breaks down the wastes and dead bodies into nutrient materials again. It may depend on other organisms for support or protection, or it may provide such services. It will have parasites, and it may be a parasite population. If it is an animal population, it will be a predator as well as a prey population.

Any community of organisms has a structure consisting of the physical distribution of its members and an age distribution of each of its populations. Thus the structure of the community is the sum of the structures of its component populations.

More important than the structure is the functioning of the community in its environment. The community and its environment comprise an *ecosystem* in which energy and materials are circulating and being transformed. The ecosystem will be some kind of a unit, identifiable from its geography and organisms and more or less separated from other units. It will never be a completely closed system, however, because energy and materials will continue to flow in and out. Even the ecosystem of the earth receives and loses energy and receives a small amount of materials.

5.2.1 CONTROLLING FACTORS

The controlling factors of ecosystems, the geological situation, the climate, and the available organisms (Fig. 5.8), are independent of the system. They are complicated sets of factors that are largely independent of each other, and many of which vary in daily or seasonal rhythms or haphazardly around mean levels. Initially, these factors determine the kinds of organisms that can survive; eventually they limit the activity and structure of the ecosystem. They are commonly modified within the ecosystem to change their effects. For example, light may be intercepted by plants and cut off from the organisms below or the composition of the bottom may be changed by burrowing animals.

The ecosystem is not a haphazard arrangement but is an ordered structure, as a result of evolution, that tends to be self-regulating. The patterns of distribution and age structures or organisms tend toward an equilibrium

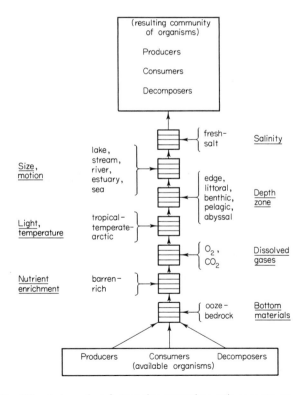

Fig. 5.8.   Independent factors that control aquatic ecosystems.

state as long as the inputs and outputs from the system remain the same. Should these change as a result of such factors as climatic change and intervention by man, the equilibrium levels will shift and trigger a new evolutionary trend toward a set of organisms that will fit better in the changed system.

5.2.2 TYPES OF AQUATIC ECOSYSTEMS

Ecosystems can be classified according to the dominant biological or physical elements of the system. Many terrestrial systems are identified by the dominant biological complex, such as tundra or tropical rain forest. Comparable aquatic systems that occur along the edges of the waters are identified as coral reefs, salt marshes, and bog lakes. Aquatic ecosystems characterized by the dominant physical factors include oligotrophic (barren) arctic lakes, tropical seas, or cold streams.

Another way of classifying the aquatic ecosystems of interest to the resource specialists but not of common use to ecologists is according to the dominant resource species and their location. Examples of such ecosystems are the following:

(a) *Oyster beds*. Subtropical to temperate estuaries in depths to about 15 m, salinity above $10\%_{00}$, bottom of gravel or rubble.

(b) *Cod and flounder grounds*. Outer continental shelves in temperate to arctic areas, gravel, sand, and mud bottoms.

(c) *Croaker grounds*. Inner continental shelves of tropical and subtropical areas.

(d) *Rockfish grounds*. Outer continental shelves and upper continental slopes of temperate to arctic waters, rock and gravel bottoms.

(e) *Tuna grounds*. Surface waters of tropical and subtropical high seas far from land around the world.

(f) *Herring and anchovy grounds*. Enriched temperate to arctic surface waters mostly near shore.

(g) *Trout streams*. Cold, well-oxygenated streams.

Such a list could be greatly extended but to no significant purpose here. Few fishery resource ecosystems have been studied intensively.

5.2.3 INTERSPECIES RELATIONSHIPS

Each species in a community influences the lives of other species and may be influenced by them in turn. The interrelationships may be mutually beneficial, as in the case of bacteria in the gut of a cellulose-eating animal, beneficial to one partner and of no great concern to the other, such as a barnacle on a whale, beneficial to one partner and harmful to the other, such as a predator and prey, or harmful to both partners, such as competition between two species for the same space or food. These relationships shape the evolution of both the species and the ecosystems as organisms adapt or perhaps become extinct. Predation and competition are of special interest to the resource scientist.

Predation.   In the popular concept, a predator is a rapacious animal, such as a tiger shark; in a more general sense it is any consumer of other organisms. It may be a herbivore, eating plants, or a carnivore, eating other animals. This broad definition includes the parasites because they live at the expense of their hosts. When a newly introduced predator population starts to consume a prey population that has been in equilibrium with its competitors and other predators, the first consequence is an increase in the mortality rate of the prey population. Removal of some prey will leave more space and food for the remaining prey, and their chances of survival and growth will improve (Fig. 5.9). If the predator population should take still more prey, the prey population would decrease further, and the compensatory survival and growth would increase more, but never enough to restore the original biomass of the prey. The predator population can make increasing use of the prey until the maximum equilibrium (sustainable) yield of prey is reached. Should prey in excess of this level of yield be taken, the excess could be only temporary and the consequence a reduced prey yield to the detriment of the predator.

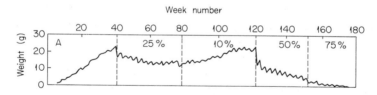

Fig. 5.9. Biomass of a population of guppies in an aquarium that was started with 5 pairs and subjected to artificial predation. After week No. 40, animals were removed at the triweekly rates shown in the panels. [After Silliman and Gutsell (1958).]

When man begins to harvest a living, wild resource, such as a stock of fish, his harvest is additional predation. He can take advantage of the resiliency of his prey population through its opportunities to increase in survival and growth, but in his own interest he cannot reduce the number in the prey population below the level of maximum sustainable yield. He can increase his benefits in the process of harvest by taking individuals whose removal will least affect the productivity of the prey, such as old, slowly growing, poorly reproducing individuals and surplus males.

On the other hand, when man seeks to control a pest either by capture or poison, he will need to overcome resiliency of the pest (prey) population and reduce its number well below the level of maximum sustainable yield. A large effort may be necessary, although the task will be easier if he can select individuals whose removal will reduce the resiliency and productivity of the pest population, such as young females during their reproductive period and larvae at a critical stage.

Competition.   Two species may interfere with each other by needing the same things at the same time and place. Usually the degree of competition is related directly to the closeness of the phylogenetic relationship among the competitors, but even diverse organisms compete with each other to some extent, e.g., a seaweed and a fish cannot occupy the same space at the same time. The consequence of competition is a change in natality or mortality rates that will result in fewer numbers of at least one of the competing populations.

The principal mode of a competition of different species is for space and food, but many other factors may be involved. Some species may degrade the environment for others through their egesta or excreta. Some may escape predators that will then seek others. Some may harbor disease organisms harmless to themselves but virulent to others. Competition may occur only during adult life, or only during larval life, or between larvae and adults.

The results of competition are difficult to observe directly in nature because of the multiplicity of competing species, the changeability of environment, and the rarity of opportunity for observation of the beginning or progress of the competitive situation. Consequently our understanding of the principles has been obtained largely from laboratory models of simplified situations. These have shown that if two extensively competing species were placed in a simple, uniform environment, one would eliminate the other, but the survivor would not always be the same one. As one species would become more abundant, it would gain competitive advantage. On the other hand, if the environment were not uniform or were changing, the two species would coexist at least for a much longer time. Each of the species would find some advantage over the other at some place or time; they would compete less. Neither species would be as numerous as it would be if it were isolated, but the two together would be more numerous than a single species in the same environment. These findings are supported by observation of natural environments. Generally, in nature more species occur in the complicated environments than in the simple ones, i.e., there are more niches to be filled; but the maximum yield of resource animals comes from the simple environments that have few species.

Predators may change the competition among prey populations. As one prey species becomes rare, the predators may shift their exploitation to a more abundant, more available species. In this case the competition between prey species will diminish and both may survive, whereas in the case where there is no predator species competition between species might lead to elimination of one. This principle is supported also by observation that in nature the coexistence of many closely related species seem to be enhanced by predators.

5.2.4 CYCLES OF ABUNDANCE

The size and structure of populations and communities are regulated, not only by the physical factors of the ecosystem, such as temperature, substrata, and depth, but also by the interplay of intraspecific competition, interspecific competition, and predation. Either the effects of the physical factors or the interplay of biological factors may lead to cycles of abundance in the populations. Abundance cycles may follow passively recurring external influences or perpetuate themselves under certain combinations of internal factors.

We shall not be concerned here with the annual cycles of the abundance of short-lived organisms that follow the seasonal changes in climate. These may be related to the changes in temperature, precipitation, or light.

The long-term changes can influence ecosystems in ways that will change greatly the balance among populations, even if they are evidenced by trend changes of only 1° or 2°C in mean annual temperature or a similarly small change in another factor. The changes may involve a major change in numbers, or in the case of some mobile marine populations, a change in location.

Self-perpetuating cycles of abundance in certain populations occur over long periods of time in some communities. The mechanisms that sustain the oscillations are the relationship of a consumer population to its food supply, in some cases at least, and the time lags in the response of a population to changes in its limiting factors. Any community in a constant environment will tend toward an equilibrium of numbers in which consumers are keeping their food supplies near the level of maximum sustainable productivity. As the consumer population increases in number, the food supply per individual decreases, and vice versa; thus the population of consumers is regulated by its food supply.

The population of consumers does not respond instantly, however, to changes in its food supply. Any increase in number must be caused by either an increase in natality or a decrease in mortality, usually of the younger organisms. The increase in biomass requires time for young to grow and perhaps time for additional generations to mature and reproduce.

Another kind of time lag occurs when limiting factors are tightening. An increase in mortality will probably affect mostly the younger organisms

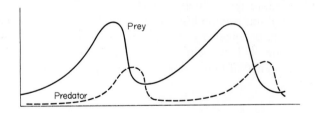

Fig. 5.10.   Diagram of a stable cycle in numbers of predator and prey.

and leave a population of adults. When the limiting factor is a food supply, the adults may exist for a time in semistarvation.

The consequence of these lags is a tendency for predator populations to either increase or decrease some time after the changes in the prey populations have taken place. When the predator and prey live in an uncomplicated ecosystem the changes in populations may be cyclic. In laboratory experiments and in a few natural populations (e.g., those of lynx and rabbit in Canada), oscillations of the kind shown diagrammatically in Fig. 5.10 have been sustained for a time. Such cycles tend to annihilate themselves when the predator population is very efficient and the prey population very vulner-

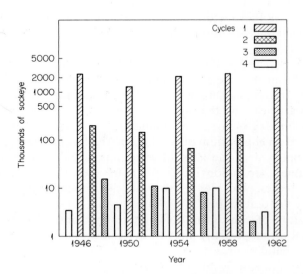

Fig. 5.11.   Annual parental populations of Adams River sockeye salmon. [After F. J. Ward and P. A. Larkin (1964). Cycle dominance in Adams River sockeye salmon. *Int. Pac. Salmon Fish. Comm. Progr. Rep.* **11**, 1–116.]

able. They tend toward a steady state when the predator population is inefficient and the prey population hard to capture. As the complications increase, however, the systems tend to have greater stability and less oscillation. Predators can turn to alternate prey, and prey can find refuges for protection from predators.

Some natural cycles in aquatic animal populations are amazingly persistent and not adequately explained. For example, the quadrennial cycles of the several runs of Fraser River (British Columbia) sockeye salmon (e.g., to the Adams River, tributary of the Fraser, Fig. 5.11) have endured for many decades. Their persistence has been attributed in part to predation by trout during the time that the young sockeye salmon live in the lakes. It was observed that, when juvenile sockeye salmon were scarce, the trout ate a large proportion of the salmon but fared poorly themselves because they could not find enough other prey. When juvenile sockeye salmon were abundant, the trout ate a smaller proportion of the salmon but fattened as they satisfied themselves. Apparently the trout population was limited to the number that could survive the years of scarce salmon.

5.2.5 SUCCESSION OF COMMUNITIES

In any new ecosystem with stable physical inputs, the community of organisms will change in composition and complexity with time. The first organisms to dominate the system will modify the environment in ways that may be detrimental to themselves and favorable to others. Organic material, both living and nonliving, will accumulate and the number of species will increase at least for a time. The trends will not be haphazard but will be directional toward a climax in which the composition and functioning of the community will be relatively stable.

Such a succession of changes is analogous in ways to the growth of individuals and populations. The succession of stages follows a predictable pattern. The young structure changes rapidly, the old changes slowly. The size, i.e., biomass of organic material, increases toward a limit. It is a biological process controlled by the community within the overall restraints of the physical environment.

The best-understood examples of succession are those in terrestrial ecosystems dominated by higher plants. The sequence of changes in a temperate climate from grassland through shrubs to pine and hardwood forests, with the associated succession of animals, has been studied extensively. Each successive dominant organism changes the environment to its own disadvantage and to the advantage of its successor. Such a sequence of changes may require more than 100 years before it reaches the relatively stable status of the mature hardwood forest.

Less well understood is the succession of aquatic populations, especially those away from the edges of waters where higher plants do not occur. In these the one-celled plants and the animals do not modify their environment in the same obvious ways that shrubs and trees do. Nevertheless, succession does occur in reasonably predictable ways, organic matter does accumulate, and an ecosystem with constant controlling factors does tend toward stability.

When the controlling factors of the ecosystems change because of either natural catastrophe or man's intervention, the patterns of succession may be reversed. The usual effect is a change in the ecosystem toward the pattern of a younger, less stable community. Often enough, when man intervenes the patterns may become essentially unpredictable.

Most wild aquatic resource populations are found in a climax state and do not provide evidence of succession, but those in which the species have changed provide examples deserving close study as possible indications of changes to be expected commonly. One of the best-documented changes is the succession of fisheries and fish species in the Great Lakes of North America.* These fisheries have been subject not only to heavy fishing but also to introductions of new fish species, pollution, and diversion of water. The same pattern of change was repeated in all of the lakes. It was seen first

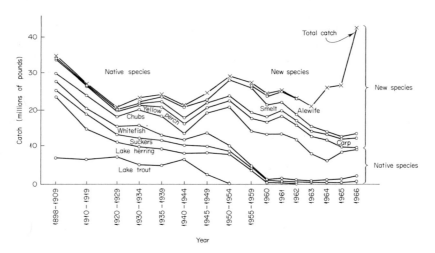

Fig. 5.12. Catches of major species in the commercial fishery of Lake Michigan (1898–1966). [Data from: S. H. Smith (1968). Species succession and fishery exploitation in the Great Lakes. *J. Fish. Res. Bd. Can.* **25**, 667–693.]

*S. H. Smith (1968). Species succession and fishery exploitation in the Great Lakes. *J. Fish. Res. Bd. Can.* **25**, 667–693.

in the lowest lake, Lake Ontario; then upstream in Lakes Erie and Huron, and finally in Lakes Michigan and Superior. Gross reduction occurred in favorite fish populations such as lake trout, lake herring, and whitefish and led to the collapse of fisheries for them. They yield of native species changed rapidly as alewives and sea lampreys invaded the lakes and as carp and smelt, introduced species, became established. The changes in the Lake Michigan fisheries were typical (Fig. 5.12). Here in the late 1960's the dominant species was the alewife. The alewife population reduced sharply the populations of other plankton-eating species. Its abundance and the nuisances caused by spring mortalities led to introduction of still other species, namely, coho salmon, chinook salmon, and steelhead trout, as climax predators. These fed on the alewife population and flourished in the late 1960's.

5.3 Material and Energy Flow

An ecosystem acquires its unity, not only because of the structure of its community of organisms, but also because of the circulation of materials and energy among members of the community. The flow of materials is cyclic; the flow of energy one way.

Certain inorganic salts, CO_2, and water are synthesized into organic materials in plants during the process of photosynthesis. The plants are

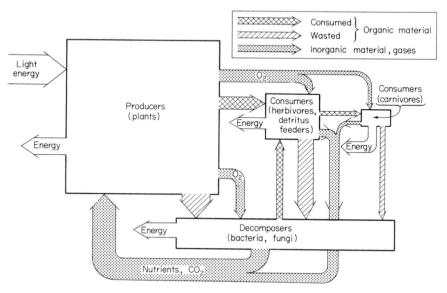

Fig. 5.13.   Diagram of the cycle of material flow in an ecosystem.

consumed by herbivores; they in turn by carnivores. The wastes produced in the process and the bodies of the animals are eventually broken down by autolysis, bacteria, and fungi into the inorganic salts, CO_2, and water that can then be reused (Fig. 5.13).

The plants that use the inorganic materials are considered to be the *producers*, the animals that eat plants or other animals the *consumers*, and the bacteria and fungi the *decomposers*. In another classification the plants are called *autotrophs* (self feeders), and the animals *heterotrophs* (other feeders). The producer and consumer stages are considered also to be trophic levels; the plants are the first level, the herbivores second, the carnivores that eat herbivores third, and other carnivores fourth or a higher level.

5.3.1 Food Chains and Webs

When a plant of trophic level one is consumed by an animal of trophic level two, and it in turn by an animal of trophic level three, the organisms form a food chain. An example of a simple, yet a major marine food chain, is the diatom–euphausid–baleen whale chain of the southern oceans. Such a clear-cut chain rarely occurs, however, because few animals belong to only a single trophic level. Many aquatic herbivores consume some bacteria and fungi in addition to unicellular plants. Many carnivores that feed normally on herbivores may occasionally eat other carnivores or plants. The common pattern of material flow through individual species of organisms is a complicated food web (Fig. 5.14).

The material cycles in the aquatic environment are not independent biological cycles but require a variety of geological cycles for their closure or augmentation. The living organisms tend to occupy a specific location,

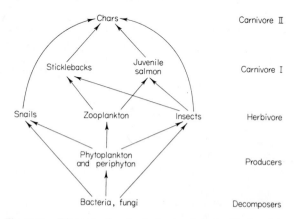

Fig. 5.14.   Diagram of a simple food web in an Alaskan lake.

but the organic particles and inorganic materials drift with the water currents or fall toward the bottom. Nutrient materials in rivers are carried out to sea and are replenished by the weathering of rocks and the ecosystems of the land. Nutrients in lakes and oceans drift with the currents and, in addition, fall toward the bottom. They are restored to the photosynthetic zone near the surface by seasonal overturns, turbulence, and upwelling (see Section 2.6.6). The material cycles are called appropriately *biogeochemical cycles*.

5.3.2 ENERGY FLOW

The material cycles receive the energy needed to maintain them from the sun's contribution of energy to photosynthesis at the producer level. The energy of the light is converted to chemical energy. The chemical energy is stored in the organic materials, principally the carbohydrates, proteins, and fats. These organic materials are sorted and partially discarded, reconstructed in the bodies of other organisms, oxidized to provide heat and energy for living, and wasted in surplus reproductive products and dead bodies as they flow through the cycle. After the energy has been captured in photosynthesis, either it is used in the maintenance of the succeeding trophic levels or it is wasted as heat. The first law of thermodynamics is satisfied: no energy is lost or gained in the cycle but is merely converted from one form to another. The second law of thermodynamics is evident throughout the process: the transformation of energy into useful work cannot be effected completely. Energy "leaks" from the cycle as heat or is "trapped" in unusable chemical compounds during all of the transformations that occur. The maximum energy is usually stored at the producer level, much less in each successive trophic level.

5.4   Production and Yield

Much of our interest in ecosystems originates in their ability to produce organic material of kinds that we want. We are concerned about the unity of the system, its efficiency, and its permanence because we want to understand it so that we can either maintain it or improve it. We may want to consider the production of an entire community in an ecosystem (e.g., a lake), a trophic level, a population, or an individual.

Our concept of the production of organic materials by living organisms has been evolving rapidly, and the student will find many definitions. Earlier in this chapter, plants were referred to as the first trophic level, or producer level. Limitation of the definition to this level, i.e., to primary production, is a usage of some authors. Another common usage is that of the word *yield* for that part of an ecosystem removed by man. Production can be used in the sense of a product as well as a process. It can be, and has been modified by the adjectives *actual, potential, net,* and *gross,* each with varied meanings. It is applicable either to the total energy or to the total materials. The student should read the literature with caution.

The preferred definition of *production* is the addition of new biomass to an organism or group of organisms in a unit of time. In the case of a single individual, production is equivalent to growth in weight plus the weight of any material lost that has been part of the body, such as a moulted shell or reproductive products. In the case of a population, trophic level, or ecosystem, production is the sum of the production of the individuals surviving plus the production of any dying during the unit of time.

Production may be expressed as weight, either wet or dry; as nitrogen content to indicate the protein fraction; or as caloric content to indicate the energy.

Production may be considered as *primary* for that part derived from photosynthesis, *secondary* for that part derived from herbivores, and *tertiary* for that part derived from the higher trophic levels. The terms *tertiary* or *fish production* may be used also to mean the production of the aquatic organisms used in part directly by man.

5.4.1 PRODUCTION EFFICIENCY

The special value of production as defined above arises during study of the transfer of materials between trophic levels in an ecosystem. Any consumer population uses part of the production of a lower trophic level and transforms part of what it uses into production. The efficiencies of this usage and transformation are of fundamental importance. The first is expressed by the ecotrophic coefficient ϵ_{n+1}, which is the fraction of the production, P, of level n that is consumed, C, by level $n + 1$; thus

$$\epsilon_{n+1} = C_{n+1}/P_n$$

The second is expressed by the utilization coefficient, K, which is the fraction of the consumption of level n that results in production by level n; thus

$$K_n = P_n/C_n$$

In a simple system with a single species at each level,

$$\epsilon_2 = C_2/P_1 \quad \text{and} \quad K_2 = P_2/C_2$$

It follows that

$$P_2 = K_2 C_2 = K_2 \epsilon_2 P_1 \quad \text{and} \quad P_3 = K_3 \epsilon_3 P_2 \cdot K_2 \epsilon_2 P_1$$

The efficiency of conversion of materials as energy from one level to the next is the product of the two coefficients, $K_n \epsilon_n$. They are frequently estimated together, and sometimes the product is called the *ecological transfer coefficient*. Actual values ranging from 0.05 to 0.3 have been estimated for $K_n \epsilon_n$ for some resource communities; the most common value has been between 0.1 and 0.2. A number of authors have used the value 0.1, however, as a useful and convenient approximation.

The photosynthetic stage of production is dependent not only upon the amount of sunlight but also on the supply of nutrient materials, temperature, and the organisms present. Only part of the sunlight is available for photosynthesis since part of it is either reflected or physically absorbed; under natural conditions, the overall efficiency is only on the order of 0.01. In consideration of such a low value and of a known efficiency of only about 0.000001 for a common, five-step food chain, it is not surprising that we try to improve the efficiency of the natural production process. A small improvement in the use of light or a reduction in the length of the food chain can change the production manyfold!

A consequence of the transfer efficiency in a steady state is that the relative biomasses of the successive levels change by a factor of about 10. This succession of changes results in a *pyramid of mass* (or *energy*) as diagrammed in Fig. 5.15.

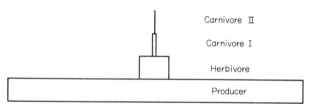

Fig. 5.15. Diagram of the relative amounts of materials or energy in four trophic levels of an ecosystem. This succession of changes is called a *pyramid of mass*, or *energy*.

5.4.2 METHODS OF MEASURING PRODUCTION

Two general methods are used to measure production although neither is entirely satisfactory. One, which can be used only for primary production, is to measure chemically the amount of photosynthesis. The other is to measure the change in biomass with time of the unit under study with an allowance for material lost or individuals dying.

The amount of photosynthesis can be estimated from either the amount of CO_2 consumed or the amount of O_2 formed. The general equation for synthesis of carbohydrates is

$$6\ CO_2\ +\ 6\ H_2O\ \xrightarrow{+\ energy}\ 6\ [CH_2O]\ +\ 6\ O_2.$$

The molecules of CO_2 used equal the molecules of O_2 produced, and the atoms of carbon used equal the atoms of carbon produced.

The measurement of CO_2 consumed at a point in the water and in time (a 4-hour period is frequently used) is accomplished by taking a sample of the water containing phytoplankton, determining the CO_2 content, adding a measured amount of radioactive $^{14}CO_2$ as a carbonate, sealing the bottle, and suspending it at the point of sampling for the test period. After retrieval of the sample the plankton is filtered and the ^{14}C determined from the radio activity. The total C uptake can be estimated then from the amount of C originally present, the amount of ^{14}C added, and the amount of ^{14}C taken up. The measurements are usually repeated in a dark bottle in order to estimate the nonphotosynthetic processes.

Alternately, the O_2 produced can be estimated from a similar sample by determining the initial O_2 content, sealing the sample in two bottles, one of which is dark, and suspending the two samples at the point of sampling. After retrieval the O_2 is determined in both bottles. The amount of O_2 produced is the increase in the transparent bottle plus the amount used in respiration in the dark bottle.

The primary production of a body of water for a period of time can be estimated by the repetition of the observations throughout the photosynthetic layer and throughout the period (including nights). The results are expressed commonly as grams of carbon (gC) produced per square meter per day, month, or year. One gC is approximately equal to 10 g of live

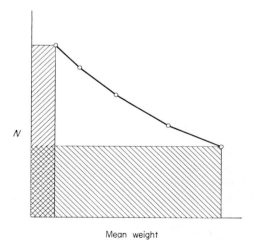

Mean weight

Fig. 5.16.   Diagram of the changes in number (N) and mean weight of a cohort at successive dates (●). The biomasses at the first and last times are indicated by the shaded areas; the production is indicated by the entire area under the curve.

weight. The range of values found is from about 0.1 gC/m²/day in barren tropical waters of the open sea to as much as 10 gC/m²/day during periods of upwelling in rich coastal waters.

When such chemical measurements are not available, the standing crop of phytoplankton can be estimated in terms of either biomass or number by towing calibrated nets and estimating the quantity per unit of area of water surface. Such a measure provides by itself a crude index of production because standing crops tend to be larger where production is greater. When it is possible to measure the change in the standing crop and the rate of loss due to grazing or death, it may be possible to calculate the production.

Secondary or tertiary production may be indicated crudely by the size of the standing crop. It may be possible, however, to identify within a higher trophic level a cohort or a set of cohorts (e.g., a year group of fish) that can be followed for a time. When estimates of the numbers and mean weight can be obtained periodically (Fig. 5.16), then it is possible to estimate the production.

REFERENCES

Most of the scientific literature on ecology has been published during the past two decades and most of the important work on aquatic ecology during the past decade. The following list includes a few general ecological references, several short introductory summaries, and some major recent works on aquatic productivity.

Andrewartha, H. G. (1961). "Introduction to the Study of Animal Populations." Methuen, London.
Andrewartha, H. G., and Birch, L. C. (1954). "The Distribution and Abundance of Animals." Univ. of Chicago Press, Chicago.
Boughey, A. S. (1968). "Ecology of Populations." Macmillan, New York.
Cox, G. A., ed. (1969). "Readings in Conservation Ecology." Appleton, New York.
Gragg, J. B., ed. (1962, 1964, 1966, 1967, 1968). "Advances in Ecological Research," Vols. 1–5. Academic Press, New York.
Edmondson, W. T., and Winberg, G. G. eds. (1971). "Methods for Assessment of Secondary Productivity in Fresh Waters," Int. Biol. Progr. Handb. No. 19. Blackwell, Oxford.
Frey, D. G., ed. (1956). Symposium on primary production. *Limmol. Oceanogr.* **1**, 71–117.
Gerking, S. D., ed. (1967). "The Biological Basis of Freshwater Fish Production." Wiley, New York.
Goldman, C. R. 1965. Primary productivity in aquatic environments. Proceedings of an I.B.P.-PF Symposium. *Mem. Inst. Ital. Idrobiol.* **18**, Suppl.
Hazen, W. E., ed. (1965). "Readings in Population and Community Ecology." Saunders, Philadelphia, Pennsylvania.
Macan, T. T. (1963). "Freshwater Ecology." Wiley, New York.
MacArthur, R. H., and Connell, J. H. (1967). "The Biology of Populations." Wiley, New York.
Odum, E. P. (1966). "Ecology." Holt, New York.
Ricker, W. E., ed. (1968). "Methods for Assessment of Fish Production in Fresh Waters," Int. Biol. Progr. Handb. No. 3. Blackwell, Oxford.

Silliman, R. P., and Gutsell, J. S. (1958). Experimental exploitation of fish populations. *U.S., Fish. Wildl. Serv., Fish. Bull.* **133**, 213–252.

Southwood, T. R. E. (1966). "Ecological Methods—with Particular Reference to the Study of Insect Populations." Methuen, London.

Steele, J. H., ed. (1970). "Marine Food Chains." Univ. of Calif. Press, Berkeley.

Tait, R. V. (1968). "Elements of Marine Ecology." Butterworth, London.

Turner, F. B., ed. (1968). Energy flow and ecological systems. *Amer. Zool.* **8**, 10–69.

Vollenweider, R. A., ed. (1969). "A Manual on Methods for Measuring Primary Production in Aquatic Enclosures," Int. Biol. Progr. Handb. No. 12. Blackwell, Oxford.

Whittaker, R. H. (1970). "Communities and Ecosystems." Macmillan, New York.

Analysis of Exploited Populations

Most fishermen, in any part of the world, claim that fishing is not as good as it used to be. They remember with more or less accuracy that, during the early years of a developing fishery, their catches were of extraordinary abundance, consisted of larger fish, and had a high proportion of fish of the more valuable species. They have seen an increase in the number of fishermen and total catches, and the development of superior, new vessels with electronic equipment and efficient new nets, and therefore despair for the fish resource.

Out of this despair have come the basic questions of fishery science. What are the causes of changes in fish populations? Are they natural fluctuations that are beyond our control? If so, can we forecast their occurrence to take maximum advantage of periods of high abundance and protect populations during periods of scarcity? Was the decline caused by fishing? If so, what amount and kind of fishing will produce the maximum sustainable yield? Or, as is more probable, were the changes caused by a combination of fishing and natural factors? If so, what is the best strategy for fishing?

The fishery scientist will approach these questions from the general viewpoint of an ecologist. He will recognize that fishing is just one of the factors limiting the numbers of fish in the waters and will try to define the populations, their niches, and the limiting factors. He will study the natural history of the fish in order to understand their migrations, reproductive habits, food, diseases, etc., but he will give special attention to the populations, their abundance, their recruitment, and their rates of mortality due to either fishing or natural causes. In other words, he will study the population dynamics and try to relate the changes in populations to the changes in fishing and to the natural factors influencing abundance.

6.1 Effects of Fishing

Let us consider an isolated, unfished pond containing fish of one species that reproduce and maintain themselves. Let us assume that the isolation and lack of fishing have existed for many years, that there are no serious epidemic diseases, and that the population is stable in abundance and age composition. The situation is a simple one of a population in a niche in which it is limited in biomass by food, space, and other factors. Each individual must compete with others in the same population for food, space, and mates.

Now let us assume that man starts to fish on the population with gear that can catch all fish over 20 cm long with equal efficiency. The result is an increase in the rate of mortality of the older fish and a replenishment of them only by growth of the fish in the group less than 20 cm long. We must expect, therefore, a reduction in abundance of the population and a reduction in the average size of the fish. This will result in more space and food for each of the remaining fish and an increase in the rates of growth and survival. These increases will not entirely compensate for the reduction in abundance from fishing; so, as we increase the amount of fishing, we must expect a continuing decline in abundance and average size. If we removed all fish 20 cm long and if these were still immature, then we would eliminate the spawners and eventually exterminate the population. On the other hand, if we leave enough spawners to supply young, then we would perpetuate the resource. We should also expect that an average amount of fishing can be found that will permit a sustained catch of maximum weight.

Such changes in a stock are indicated by the ecological principles summarized in the preceding chapter and are known to occur actually in nature, but they are obscured frequently by the practices of fishermen. Fishermen fish for the most valuable combinations of species and sizes on the nearest grounds. As the fish on these grounds become scarcer, either with the seasons or years, fishermen move to more distant grounds or seek other species. Moreover, they learn to locate more easily the fish in particular areas and frequently improve their nets, vessels, or navigational equipment in order to catch fish more efficiently. These changes, or shifts, may obscure the characteristic reduction in abundance and size of fish in a single population.

When a new fishery develops as a consequence of the discovery of a new stock and the promotion of suitable markets, the production trend is frequently exponential (Fig. 6.1) in the early stages, but then decline may set in. As money is made, more investors are encouraged, and more production is attained. The potential seems exciting because the fishery is harvesting an accumulated stock that has not been exploited previously. Usually, the excitement is sustained by the discovery of new grounds, better gear, and improved fishing strategy. The actual sustainable productivity of the resource is

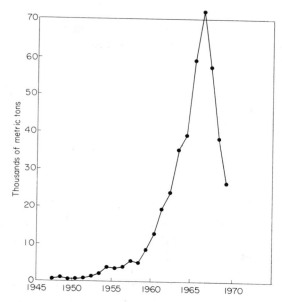

Fig. 6.1. The trends in U.S. production of Alaskan king crab from 1947 to 1969 are an example of production trends in a new fishery. [Data from statistical publications of the Alaska Department of Fish and Game.]

commonly unknown and difficult to estimate (note the assumptions of equilibrium conditions in the yield models, Section 6.9). The actual changes in the stock are obscured, not only by the practices of the fishermen (noted in the paragraph above), but also by the natural fluctuations and by the time lag required for the stock to approach equilibrium. Too much effort is applied and the unhappy consequence in many fisheries has been a catch trend that overshot the level of sustainable yield. This has been followed by a decline of sustained productions to levels ranging from one-fifth to one-third of the peak production reached during the development stages. The tendency to overshoot the sustainable level of yield is greater among long lived species such as halibut and king crab than among short lived species such as anchovy and penaeid shrimp.

6.2 Stocks and Migrations

Let us assume that fish of one species are being caught by a fleet operating on a fishing bank that is small enough so that the fish intermingle extensively during the reproductive cycle. Let us assume further that each individual has an equal chance of mating with any other individual of the same species and opposite sex on the bank; i.e., the fish are a fully interbreeding group.

With the fishing on this bank is enough to affect the abundance and average size of the remaining fish, i.e., the success of fishing, where else is the success of fishing affected? Would fishing for bass at one end of a 10-km-long lake affect the success of fishing at the other end? Would fishing for albacore off Japan affect the fishing for albacore off California?

The answers to such questions would depend on the extent that one group intermingles with others. Should the intermingling be extensive, then all the intermingling groups would be affected by fishing in an area occupied by any group. Should the fishing on one group need to be controlled, then fishing would need to be controlled in all areas occupied by that group.

To an ecologist, determining the extent of intermingling of such groups is a problem of determining the niche of the catchable parts of the populations. He may also consider the problem as that of determining the extent of reproductive separation of the populations. To a taxonomist, the problem is one of determining the population structure of the species. To a fishery scientist, this step is one toward defining the stocks with which he must deal.

The student will find in this subject a bewildering array of semantic problems because there is little agreement on the meaning of the words used to define groups in the hierarchy with the rank of subspecies and below (see the discussion on nomenclature, Section 3.1). The rules of international nomenclature recognize *subspecies* as the least inclusive category, but within that category the terms *population, subpopulation, stock, race, variety, breed, strain*, as well as others, are in current use. Moreover, some of these words are used for population units with legal or other nonbiological implications. For example, *stock* has been defined in some international treaties as a group of fish capable of independent exploitation and management.

Further complications arise because most species and subspecies are composed of local populations, no two of which have identical genetic characteristics. The populations are of one or more of the following categories: (1) populations that are geographically separated from each other, with slight opportunity for genetic interchange; (2) a series of gradually changing, contiguous populations, called *clines*; (3) a set of sharply differentiated contiguous populations with zones of hybridization in between (*stepped clines*).

In this volume, heretofore, the words *population* and *stock* have been used without exact definitions, so some definitions are in order. *Population* is used to mean a biological unit and *stock* is used to mean an exploited or management unit. It will be obvious that the ideal definition of *stock* is that of a single interbreeding population, but this condition is so rarely demonstrable, either because of scanty data or because of the rarity of isolated interbreeding populations, that *stock* must be more or less arbitrarily defined. It is a unit capable of independent exploitation or management and containing as much of an interbreeding unit or as few reproductively isolated units as possible. As such, it may include several species, e.g., the stock of antarctic whales or the stock of several species of salmon that intermingle in a fishery. Thus, by usage in this volume a *stock* is a management unit defined for operational purposes, whereas a *population* is the actual biological unit and may never be completely described.

6.2.1 MIGRATIONS

Migration is an adaptation of many aquatic animals to increased abundance in a varied environment. It enables them to use rich feeding areas, to protect eggs and larvae as may be necessary, and to avoid unfavorable temperatures, salinities, or other changes in aquatic climate. Some animals can find all of their requirements in one location, but most species migrate at least for short distances for feeding, spawning, or overwintering. Some may make transoceanic migrations, such as certain whales, tunas, and salmon. The value of the adaptation for increasing abundance is illustrated by the tendency among closely related species for the migratory kinds to be most abundant, e.g., the Pacific salmons are more abundant than the trouts. Most of the major food fishes, crustaceans, and mammals are migratory.

Migrations are accomplished either by passive drift or by active swimming. The eggs or larvae of many marine species and a few freshwater species are pelagic and may drift for hundreds of kilometers. When they do, then a migratory cycle is completed by the return migration of either juveniles or adults so spawning will recur in the same place.

Understanding the diverse migratory cycles of aquatic animals poses some major biological challenges. Usually the cycles occur according to the season, and the periods of active swimming follow periods of fattening or maturation of sexual products. The active migration must be a response, either imprinted or inherited, to some kind of guidance information. The simpler migrations may be accomplished by random searching or by following a gradient of depth, temperature, or salinity. Some fish follow a current gradient upstream or downstream. Some find and follow the distinctive odors of certain streams through highly developed olfactory systems. Those that migrate on the high seas perform remarkable feats of navigation by methods about which we can only speculate at this time. They may receive guidance either from the sun, from the magnetic field of the earth, or from current systems of the ocean. Finding out about these migrations is important for the utilization of the stocks, as well as for the solution of challenging biological questions, and probably will receive a great deal of attention in the future.

6.2.2 INTERMINGLING

Determining either the extent of intermingling during the course of migration or the degree of reproductive separation, the converse, is a difficult task because we cannot recognize individual animals throughout their lives; we must rely instead on a variety of circumstantial evidence. The best evidence is furnished by attaching tags to animals and recapturing them later. Evidence of migration from marked animals will establish that intermingling is occurring but may not indicate the amount because tagged fish can be caught rarely more than once and the fishing is never distributed evenly over the animals' range. Other evidence may be sought from studies of larval drift; from correlations in population characteristics, such as abundance, age composition, or incidence of parasites; from structural similarities, such as shape or number of body parts; and from chemical similarities, such as antigens or resistance to disease. Any evidence of these kinds may be useful to the fishery scientist who must make decisions about the geographic limits of stocks.

Population Characteristics.   Fishermen have learned the migratory habits of many juvenile and adult aquatic animals by accident. They try to operate in areas where animals congregate either to feed or to spawn and avoid areas where animals are scarce or absent. Sometimes they can follow migratory groups, e.g., the mackerel on their northward migration off Eastern North America (Fig. 6.2). Always they watch for evidence of migration in the changing relation between catches and time, current, temperature, or catches elsewhere because their success depends largely on being at the right place at the right time.

The biologist, too, can learn much from an analysis of catches according to time and location and an apparent shift in centers of abundance. When, in addition, observations on age composition, sexual condition, and food habits are available, then he can make inferences about the age groups involved and the reasons for the migrations.

A distinguishing characteristic of certain populations is the presence of unique parasites. When these are acquired by young animals and carried throughout life they are good marks. For example, a large proportion of the sockeye salmon from the Bristol Bay area of Alaska carry *Triaenophorus*, a cestode parasite of the muscles that does not occur in any other sockeye salmon. *Dacnitis*, a nematode of the intestine, occurs in a small proportion of the sockeye salmon of the Kamchatka area and nowhere else. The occurrence of these parasites in sockeye salmon taken at sea is a useful indication of the origin of the individuals (Fig. 6.3).

Marks and Tags.   Hundreds of ingenious ways of marking aquatic animals for later identification have been devised. All have shortcomings, so

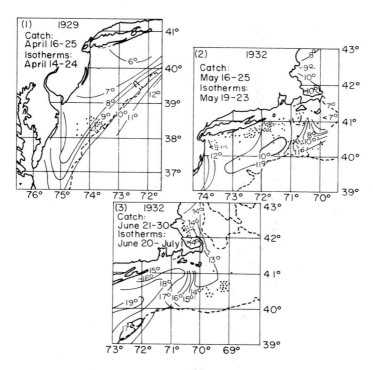

Fig. 6.2.   Distribution of mackerel catches (shown by dots) at intervals during the spring in relation to surface temperatures (°C). [Adapted from O. E. Sette (1950). Biology of the Atlantic mackerel (*Scomber scombrus*) of North America. Part II. Migrations and habits. *U.S., Fish Wildl. Serv., Fish. Bull.* **51**, 249–358.]

the fishery scientist must be familiar with many techniques and the problems of using them. Ideally the mark should be applicable to the animal without injury and with no subsequent effect on the rate of growth, rate of maturity, behavior, and liability to capture either by fishing gear or by predators. Marks fall into two categories, those that will identify a few groups, or batches, and those that will identify thousands of individuals. The former are commonly known as *marks*, the latter usually as *tags*.

Fish marking includes mutilation such as clipping minor fins, punching holes in fins or opercle, branding with heat or cold, tatooing, dyeing with vital stains, and spraying with fluorescent plastic particles. In some cases dyes are placed in the food that color the flesh, or substances are fed that are deposited either in scales or in otoliths. Generally, marks are used on animals that are too small to carry a tag of suitable size and on batches of large numbers of animals, e.g., several hundred thousand.

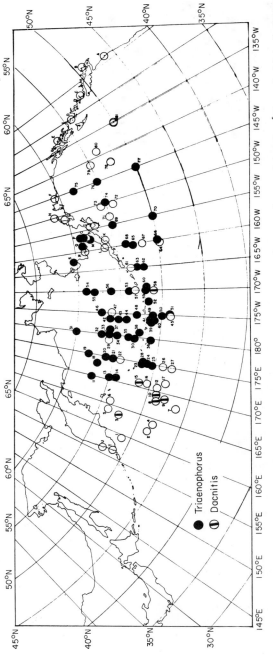

Fig. 6.3. Distribution of sockeye salmon samples taken in 1959 and occurrence of *Triaenophorus* and *Dacnitis*. [From: L. Margolis (1963). Parasites as indicators of the geographical origin of sockeye salmon, *Oncorhynchus nerka* (Walbaum), occurring in the North Pacific Ocean and adjacent seas. *Int. N. Pac. Fish. Comm., Bull.* **11**, 1–156.]

Almost all marks and marking procedures injure the fish in some way. The amputation of fins causes a slower rate of growth. Handling leads to some injuries, especially in the case of wild fish. Further, few marks are recognizable by the public, so identification at the time of recapture must be accomplished by trained people. Despite these difficulties, marks have been used extensively to supply essential data on the migrations and survival of hatchery-reared salmonid fish.

Tags that are large enough to carry a number have a great advantage in providing individual identification. Those that are external tags with instructions for return can be found and returned by the public and thus do not necessitate an inspection program. Such tags should be permanent, noncorrosive, and nontoxic. They are usually made of metal such as stainless steel, silver, or nickel, or of plastic (Fig. 6.4). One of those devised early, and still one of the more useful is the Petersen tag; it consists of a pair of discs joined by a metal pin passed through the body of the fish. Other tags in common use are made of spaghettilike plastic tubing that can be passed through the body and tied in a loop or attached to an anchor imbedded in the muscle.

Especially useful for small, delicate fish are small strips of either metal or plastic that are placed internally. Most of the metal ones are magnetized or made radioactive and are detected when the fish are passed through the field of an electronic detector. Such tags have been used for large-scale experiments on fish, such as herring, that are handled largely by processing machinery to which a detector can be added.

Tags on large animals, such as whales or large tunas, must be applied usually without removal of the animals from the water. They are delivered either from a gun or on the head of a harpoon. Usually such tags consist of an anchor that imbeds in the musculature and a colorful plastic trailer that serves to call attention to the tag.

Tags on crustaceans are lost when the animals moult unless they are attached to the musculature near the isthmus at which the shell splits during moult. Such tags have been devised for blue crab and king crab (Fig. 6.5).

Shelled mollusks that cannot be tagged are marked by filing a place on the shell until it is smooth and numbering it with a special, quick-drying paint. The mark must be waterproof and must last a long time, and few paints are satisfactory. Batches of mollusks may be marked by mutilation of the shell with either a file or an awl. Larval mollusks may be marked in batches with dyes that will remain as a ring in the shell.

The information obtained from tagging and marking programs is used in two general ways: to provide evidence of migration and intermingling, the topics of this section; and to provide information on growth and mortality rates (see Sections 6.7 and 6.8). For either use it is important that the tags

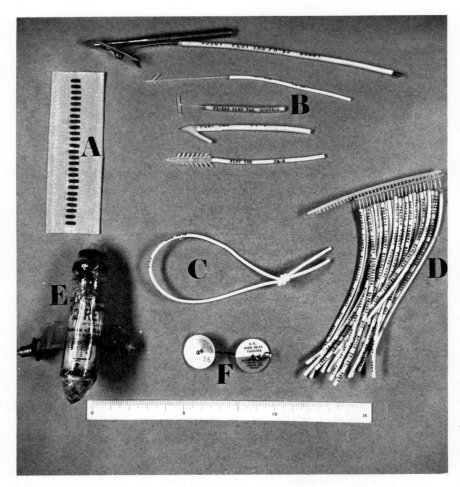

Fig. 6.4. An assortment of modern fish tags. (A) Internal plastic tags suitable for fish about 10 cm long or for small crustaceans. (B) A number of dart tags. The largest is shown attached to the metal tip of a small harpoon which is used to insert the tag in a tuna or spearfish. All are inserted into the musculature of the back. (C) A spaghetti tag of flexible plastic that is inserted through the back of a fish near the dorsal fin. (D) A partially used "clip" of dart tags that are inserted with an automatic tag applicator. (E) A sonic tag that can be attached to the back of a fish and followed with acoustical instruments. (F) A Petersen tag of two plastic discs joined by a stainless steel pin inserted through the back of a fish. (Photo by W. F. Royce, University of Washington, Seattle, Washington.)

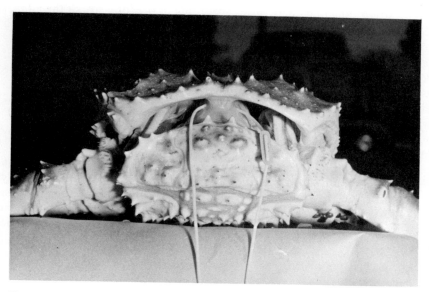

Fig. 6.5. A spaghetti tag threaded through the musculature of a king crab such that it will be retained throughout the moult. (Photo courtesy of Prof. K. K. Chew, University of Washington, Seattle, Washington.)

remain attached or that the marks remain recognizable and that the fish not be handicapped for the duration of the experiment. Any loss of tags, mortality, or change in catchability of the fish must be measured. In addition, the tag or mark reporting must be adequate and the rate of overlooking or non-reporting must be known. Various advertising and reward systems are used to encourage the return of tags.

Structural Characteristics. There is a tendency toward similarity among individuals of any isolated interbreeding population because of exchange of genetic material and a tendency in the same population toward difference from other populations of the same species because of evolution in response to different environmental restraints. The differences may not warrant designation of separate subspecies but may suffice to separate the individuals of one population from those of other populations. Fishermen frequently recognize stocks from subtle differences in color or in shape of the body. Biologists have quantified differences as *morphological* when they relate to size of body parts and *meristic* when they refer to number of body parts, such as scales, fin rays, or vertebrae. Differences of color in tone or intensity are difficult to quantify, but differences in distribution of color are easy to measure occasionally.

The size of any body part varies according to the total size of the animal together with an allometric growth factor (or change in body shape with size) and random variability within the population. A comparison of body parts between populations is made by measurement of the three kinds of variability and their difference. Sometimes a part will grow at the same rate as the whole body (isometrically); then the relative size of the part may be expressed as a ratio, or a percent, or, more commonly, a per mille. When, as frequently occurs, the part grows allometrically, then the ratio changes according to the size of the fish and one determines the relationship by regression* analysis.

An example of allometric growth in a fin of yellowfin tuna is shown in Fig. 6.6. The graph shows that this fin grows at a different rate than the body does, and an average ratio of fin length to body length is misleading at most lengths. This trouble was avoided in the analysis of the data on yellowfin tuna by transformation of the data to logarithms so that the data could be brought approximately to a straight-line relationship and the fitting of

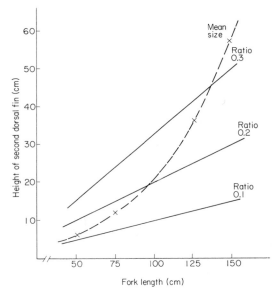

Fig. 6.6. Growth of the second dorsal fin of yellowfin tuna, an example of allometric growth. [Data from W. F. Royce (1964). A morphometric study of yellowfin tuna, *Thunnus albacares* (Bonnaterre). *U.S., Fish Wildl. Serv., Fish. Bull.* **63**, 395–443.]

Regression is the statistical term for a functional relationship of one variable on another—in this case the size of the body part on the size of the fish. In this and the ensuing discussion the student should consult a statistical text.

regression lines to the data separately for each of three segments, i.e., small, medium, and large. Then the size of the part was calculated from the regression for the central length in each segment.

The number of body parts, such as vertebrae, scales, fin spines or rays, gill rakers, and pyloric caecae, may be counted after careful specification of what is or is not to be included in the count. This step is especially important because the differences in counts between populations may be small and may be masked if different counters interpret fused vertebrae, branched fin rays, buried gill rakers, etc., in different ways. Counts change with the size of the fish only occasionally.

After either morphometric data or meristic data have been obtained for samples from two or more populations, the means may be compared. They may be compared by a statistical test of significance, i.e., a computation to determine the chances of the means being drawn from a single population. Chances of one in twenty or less are usually taken to indicate that the means differ significantly and are probably from different populations. The test results depend, however, on the number of observations in the sample; the greater the number, the more likely that real differences will be detected. Furthermore, natural populations are so variable that real differences may be expected to be found routinely when samples are large enough. Under these circumstances the test for statistically significant differences will lead only to trivial conclusions.

Of more worth and in extensive use by taxonomists is the concept of overlap of frequency distributions (not to be confused with geographic overlap). When the counts of a character from two samples are plotted, commonly they show differences (Fig. 6.7). These are expressed frequently in terms of the mean, twice the standard error of the mean, standard deviation, and range. Commonly, these are plotted in a form that shows at a glance the nature of the distribution. When the black bars just meet, the odds against the difference between means occurring by chance are about twenty to one. When the hollow bars just meet, the overlap is 16%. In the example given, the difference is highly significant statistically, but the overlap is greater than 16%.

Similar statistics for comparison of the size of body parts may be obtained from regression analysis which takes into account the effects of changes in total body length. Instead of an ordinary mean, the mean value is estimated from the regression for body length to be used for comparison. Instead of the standard error of the mean and the standard deviation, similar measures of the distribution around the line may be used.

In practice, a comparison of two samples by use of the overlap of a single character is rarely satisfactory because organisms that differ in one character differ also in others. The number of possible characters for

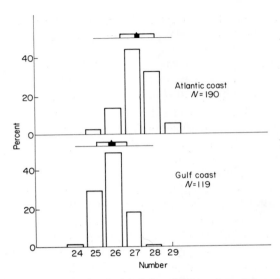

Fig. 6.7. Percent distribution of articulated dorsal fin rays in samples of weakfish, *Cynoscion regalis*. The diagrams show a small triangle for the mean, a black bar for twice the standard error of the mean on either side of the mean, a hollow bar for the standard deviation on either side of the mean, and a base line for the range. [Data from I. Ginsburg (1953). The taxonomic status and nomenclature of some Atlantic and Pacific populations of yellowfin and bluefin tunas. *Copeia* **53**, 1–10. Graphical method from C. L. Hubbs and C. Hubbs (1953). An improved graphical analysis and comparison of samples. *Syst. Zool.* **2**, 49–56 and 92.]

comparison is huge, but using more than one is statistically difficult. A great variety of sums, ratios, and products have been used. Comparisons have been attempted with coefficients of racial likeness, multiplefactor analysis, and many other methods. The most satisfactory mathematical approach is based on a measure of difference between means in units of the standard deviation. It allows consideration of multiple characters by adding to the difference between means for the second and additional characters only to the extent that they are not correlated with characters previously considered.*

Some tagging and morphometric studies of fish populations have shown that fish migrate from one location to another but that fish in the two locations show morphological differences. These two findings can be interpreted

---

*A full treatment of this mathematically complicated problem is beyond the scope of this book. For a discussion of an application to fish populations, see W. F. Royce (1957). Statistical comparison of morphological data. *In* J. C. Marr, Contributions to the study of subpopulations of fishes. *U.S., Fish Wildl. Serv., Spec. Sci. Rep.—Fish.* **208**, 7–28; and for an introduction to the mathematics, see C. R. Rao (1952). "Advanced Statistical Methods in Biometric Research." Wiley, New York.

as supplementary estimates of intermingling. The migrations demonstrated by the tag returns show that intermingling is occurring; the morphological differences show that intermingling is not complete. When the morphological comparison includes consideration of several characters, then the overlap may provide an estimate of the maximum amount of intermingling that could have occurred.

Biochemical Differences. Genetic differences between closely related populations may appear as biochemical differences as well as structural differences. Some differences, presumably biochemical, may be detectable as behavioral differences, as tolerance to environmental factors, or as resistance to disease. Other differences may be detectable by direct biochemical tests of the composition of the blood or other tissues.

Comparison of populations on the basis of behavior, tolerance of environmental factors, or resistance to disease must be done with living animals. For this reason such studies are made on domesticated animals more commonly than on wild populations. Further, they are made frequently for the purpose of selective breeding to isolate and preserve characteristics that make the animals more amenable to aquacultural practices. Such comparisons have led to the discovery of races of oysters that are resistant to the Malpeque disease, trout that are resistant to furunculosis, and trout that grow better because they are less aggressive or tolerate better certain water conditions. A few wild populations have shown behavioral differences, e.g., sockeye salmon fry that migrate upstream on emergence from the gravel instead of the usual downstream movement.

Comparison of wild aquatic populations by the use of biochemical methods has advanced rapidly during the past two decades with the development of extrasensitive methods of separating organic compounds and the application of immunochemical methods to the problems. These methods have been applied in comparisons of a number of body tissues and seem especially promising in the case of blood because the differences in blood types among some animals have been shown to be genetically controlled.

The serological comparison of blood depends on the presence of antigens (substances, such as proteins, that cause the production of antibodies) either in the cells or in the sera. When antigens are sought in the blood cells of an animal, these are separated from the sera and injected in a laboratory animal, e.g., a rabbit, that will produce antibodies. When the antigens are sought in the sera, this is injected separately. The laboratory animal may be entirely different from the animals being tested, but more sensitive tests are usually possible with laboratory animals of the same species as those being tested. After some time and perhaps several injections, the laboratory animal may produce antibodies. When brought together with the blood of an animal similar to the original antigen producer, these antibodies will agglutinate the cells or form a precipitate with the serum.

The technique of comparing cell antigens is to bring together the cells and the serum containing antibodies in a test tube and measure the amount of agglutination (Fig. 6.8a). Usually sera containing antigens are tested against a serum containing antibodies by being allowed to diffuse through agar on Ouchterlony plates (Fig. 6.8b). The presence and amount of precipitation is measured. Sometimes the antibodies in a serum can be partly agglutinated and the remainder can be used for additional tests. Such tests are extremely delicate and require great care in the handling of the samples to avoid either chemical changes or bacterial contamination.

In addition to serum containing antibodies produced by laboratory animals, ordinary sera from any animal or plant may be used in the search for antigenic differences. The chances of such sera giving a reaction to one

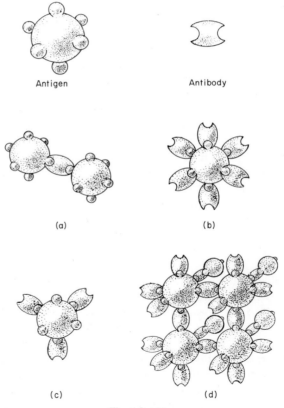

Antigen   Antibody

(a)   (b)

(c)   (d)

Fig. 6.8. (A)

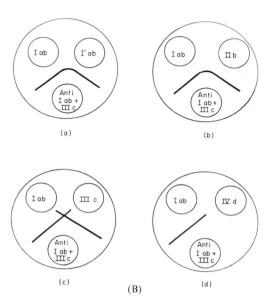

Fig. 6.8. (A) Diagrammatic representation of the lattice theory of antigen–antibody reactions. (a) Soluble primary aggregate formed in excess antigen. (b) Primary aggregate formed in excess antibody. (c) Primary aggregate formed at zone of optimal proportions. (d) Insoluble lattice formed by interaction of the primary aggregates shown in (c). (B) Types of comparative reactions in agar-plate diffusion method of precipitin analysis. (a) Reaction of identity, complete fusion of line. (b) Reaction of partial identity, "spur" formation, or partial fusion of line. (c) Reaction of nonidentity, crossing of lines. (d) Lack of reaction, absence of line. [After G. J. Ridgeway, G. W. Klontz, and C. Matsumoto (1962). Intraspecific differences in serum antigens of red salmon demonstrated by immuno-chemical methods. *Int. N. Pac. Fish. Comm. Bull.* **8**, 1–13.]

population of a species and not to another are very small, but such comparisons are so easy that thousands can be tried. Limited work on fish has produced some encouraging results. It has consisted of a comparison of skipjack populations by the use of bovine sera and a comparison of sockeye salmon populations by the use of pig sera.

The special usefulness of blood typing for recognizing interbreeding populations arises from the occurrence as alleles of certain blood types in populations. Such blood types are revealed by the presence or absence of certain antigens that are maintained in constant proportions in interbreeding populations, subject only to mutation and random variation. This phenomenon in its general sense is known as the *Hardy-Weinberg law*. It permits a comparison and judgment about interbreeding of populations when samples are large enough to permit the establishment of frequencies with sufficient limits of accuracy.

Some important studies of this kind have been made on sockeye salmon in the North Pacific.* Several samples of the sockeye salmon's sera were compared on Ouchterlony plates with serum containing antibodies produced by a rabbit after injection with serum of sockeye salmon from Cultus Lake, British Columbia. The comparison produced 14 lines of precipitation by antigens, of which 2 were missing in 8 samples of sockeye salmon from the Okhotsk Sea but were present in 10 of 11 samples from the Naknek River, Alaska. More extensive sampling and tests showed that 95.8% of the sockeye salmon from the American areas had either antigen I or antigen II and 92.1% of the Asian sockeye salmon lacked these antigens. These comparisons were extended to samples from the high seas and used in an attempt to show area and time of intermingling.

A continuing effort is being made to identify members of populations by means of highly sensitive chemical techniques. Comparisons of muscle or blood proteins by electrophoretic methods and those of elemental composition by γ spectrum analysis have shown promise.

6.3 Collection of Basic Data

After defining a stock that is being fished, the fishery scientist will seek to measure the effects of fishing on the stock. He can do this only by correlating the changes in amount of fishing with the changes in growth, distribution, reproduction, and especially, total mortality rate. The last is measurable indirectly either through marking experiments or from changes in either catch per unit of effort, or age composition, or total size of the stock.

The bases for such correlations are accurate and consistent data that are obtainable only from suitable, continuing statistical systems. The fishery scientist will want data on the amount of fishing according to time and unit area; the catch in weight or number by species, time, and area; and the size, sex, and age composition by species, time, and area. The area divisions must be chosen to agree with the known or probable stock locations and sometimes must be designated by depth also. The time divisions should agree with season, fishing periods, fish habits, etc. Usually days, months, and years are suitable units.

Such scientific data may be useful to the fish trade, especially when additional data are available on price and value, employment, amount and value of processed products, kinds of craft and equipment used, quantities

---

*G. J. Ridgeway, G. W. Klontz, and C. Matsumoto (1962). Intraspecific differences in serum antigens of red salmon demonstrated by immunochemical methods. *Bull. Int. N. Pac. Fish. Comm.* No. 8, pp. 1–13.

in storage, and exports and imports. When the fishery is recreational, then data on the number of people, their expenditures, their time spent, and the equipment used are useful to the trade that supplies services and equipment. In addition, many of these data provide an essential basis for governmental decisions about food supplies, port facilities, water usage, and taxation.

The statistical systems for collecting scientific data on the effects of fishing on the stocks will be combined, therefore, with systems used for other purposes. The fishery scientist will use as much data as possible from the general statistical system and design a subsystem to collect supplementary biological data. Usually for food fisheries, the general system will supply complete data on catch according to species, time, general location, and weight and on the amount of fishing by time. Additional information on the exact location, time, and depth of fishing, on the catch by area, on the size, sex, and on the age composition by species, time, and area will be obtained by a special system, usually a sampling system. For recreational fisheries any general system is likely to provide only data on catch by species, time, and general location. Such systems are relatively inaccurate and need to be checked as well as supplemented by carefully designed sampling systems.

Any fishery statistical system requires carefully defined objectives, supervision to insure the continuity of comparable data, and cooperation of the people who supply the statistics. Usually, one achieves cooperation by making it easy to supply the data, by keeping personal data confidential, and by making the use of the data known to those concerned.

6.3.1 THE GENERAL SYSTEM OF FISHERY STATISTICS

The general fishery statistical system is usually the responsibility of the fishery agency rather than of a general statistical agency. The system is frequently a costly part of the fishery agency's activity. It is an activity that deserves the close attention of fishery scientists; also, because the analysis of the effect of fishing depends on the reliability of the basic statistics. These are collected in ways that fit the practices of the fish trade. A number of systems are in common use.

Record of First Sale. Almost all food fish are sold by fishermen rather than bartered and almost always fishermen and/or buyers keep a record of the sale. In many countries a third copy of this first sale is required by law to be furnished to the fishery statistical system. Normally the sales record shows the names of seller and buyer and the amount and value of the sale by trade categories. In addition, the record may contain the name of the vessel, the kind of gear, and the area where the catch was taken. The state

STATE OF WASHINGTON FISH RECEIVING TICKET **No. S 0602**

DEALER'S NAME **DAHL FISH CO.** STATION

RECEIVED FROM:
NAME OF FISHERMAN OR FIRM: ADDRESS:

BOAT NAME: PLATE NO.

PLACE CAUGHT: GEAR USED: DATE OF LANDING 19

DEALER – DO NOT WRITE IN THIS SPACE

DIST.	DEALER	PORT	BOAT	TICKET NUMBER	GEAR	CATCH AREA
	200					

SPECIES AND DESCRIPTION	UNITS EFFORTS	SPECIES CODE	POUNDS	NO. OF FISH	PRICE	AMOUNT
DOVER SOLE		205				
ENGLISH SOLE		206				
PETRALE SOLE		207				
ROCK SOLE		209				
LING COD		231				
TRUE COD		241				
SABLEFISH		221				
ROCKFISH		251				
PACIFIC OCEAN PERCH		254				
FLOUNDER		212				
					TOTAL	

(DEALER—DO NOT WRITE IN THIS SPACE)

DEALER'S RECEIVER SIGNATURE

I Certify that these Fish or Shellfish were taken: (Check One) Inside Territorial Waters ☐; Outside Territorial Waters ☐, and that all other information on this ticket is true and correct

FISHERMAN'S SIGNATURE DEALER'S COPY 60

Fig. 6.9. A fish sales slip used by the State of Washington for gathering fishery statistics. One copy each goes to the dealer, the fisherman, and the state.

usually furnishes free of charge the blank forms showing the information desired (Fig. 6.9). This system is probably the best that has been devised yet to provide routinely accurate data on the catches.

Records of Fishing. Many commercial fishermen keep personal records of their catches and fishing activities either in a vessel log or a business log. These may provide invaluable information to the fishery scientist on location and amount of catches and amount of fishing by area. Special log books showing the information desired are supplied free of charge to fishermen by some fishery research organizations. The information is copied by the research organizations and the log book is returned to the fisherman.

A personal log for the season is required in some recreational fisheries. This may be a punch card in which the fisherman notes the capture of a fish by punching a hole and adds the data and location. This log is turned in at the end of the season.

Interviews with Fishermen. When large numbers of fishermen land their catches in one location, it may be possible for an interviewer to talk with either a sample or all of the fishermen as they land to obtain details of their catches, gear, and location of fishery for each trip. In addition the interviewer may be able to perform *market sampling*, i.e., to examine, measure, and weigh catches, and to collect scales and other data. Such a system may provide a valuable supplement to the records of catches obtained from either sales records or log books.

In recreational fisheries where fishermen are scattered in boats or along streams, they may be visited in the field and their catches examined and counted. This procedure is called a *creel census*. Frequently it will provide the best possible data on recreational catches, but the records must be adjusted to allow for the catches of fishermen who are not interviewed and the catches made by fishermen after they have been interviewed.

In recreational areas where access can be controlled, fishermen may be required to show their catches when leaving.

Periodic Canvass. In places where daily records are not obtainable, the only alternative may be an occasional canvass. Village leaders may be asked to estimate the seasonal catches by the village. Fish buyers may be asked to estimate their total purchases for the season. A sample of fishermen may be asked to estimate their annual catches. Such a method may produce statistics with large errors, but it may be still the best possible.

Related Statistics. A broad variety of statistics may be related to the fish catches by some kind of a ratio. Freight records or export records may be related to total catches by an estimate of the fraction shipped. Boat rental records of a recreational fishery may be combined with records of catch per unit for estimation of totals. Village census data may be combined with estimates of catch per family. Either fishermen or vessel licensing records may be combined with estimates of catch per person or per vessel. Counts of vessels made from the air may be combined with estimates of the catch per unit. Such data may be the only means of estimating catches or provide useful checks on other methods.

6.3.2 THE SUBSYSTEM OF BIOLOGICAL STATISTICS

Usually, the general system of catch statistics will not provide sufficient information on species, area of catch, size composition, or sex composition for the fishery scientist who needs to determine the effect of fishing. He will use the general system as a statistical base and obtain the additional data from catches by a research vessel or from samples of the catch by the fishery.

Sampling presents special difficulties for fishery scientists. When the samples are to be used as the basis for valid statistical inferences about the (statistical) populations from which they have been drawn, samples must be representative, that is, they must have been obtained by a method that insures that the characteristics of all possible samples drawn by the method will bear a known relation to the corresponding characteristics of the population being sampled.* When the samples meet this requirement, the individuals in them must have been chosen by some type of *random selection*, possibly by simple *random sampling* or a modification, such as stratified random sampling.

Random sampling in the statistical sense is an exact concept and does not mean haphazard selection of samples. It means that every member of the population being sampled has an equal chance of appearing in the sample. Some statisticians argue that a random sample cannot be drawn consciously but that some mechanical randomization (such as dice, random-number tables, etc.) must be used. They point out that if samples were just grabbed, even if every individual had an equal chance of entering a sample, the individuals in the *grab samples* would tend to resemble each other more than individuals in true random samples; consequently, the estimates of variability in the grab samples would overestimate the significance of the difference between the means of the grab samples and underestimate the variability in the population.

If the samples are to have the characteristics of a sampled population, the latter must be specified with care. Usually a sample may be considered to have been drawn from any of several populations. For example, a biologist samples fish in a bin on the deck of a boat. He might consider his sample to be from the population in that bin, from the catch of the net that filled his bin among other bins, from the day's or month's catch of the vessel, from the catches of the fleet, from the population of fish on the bank, or from the population of fish along the whole coast. He must specify the *sampled population* with its units and probably will do so with a *target population* in mind to which he would like his conclusions to apply. A common circumstance is to specify the sampled population as the catch of a fleet from a fishing ground during a period in order to gain knowledge of the target population consisting of the stock of fish on those fishing grounds.

Sampling programs should be designed for achievement of the best balance between accuracy of results and cost. Accuracy of results depends in general on two factors: *variance* of the estimated mean and *bias*, or systematic error. Variance of the mean is caused by random errors and is reducible either by procurement of larger samples, by sampling of strata more efficiently, or by sampling in stages. Bias may be much more serious, however, because rarely can one examine whole aquatic populations to provide checks on sampling procedures, and agreement among repeated samples is no guarantee of a lack of bias. The problems in fishery sampling differ in character, depending on whether the information sought concerns the structure of the population or its abundance.

---

*The student who is not familiar with sampling theory should consult a modern statistical text.

Sampling to determine structure, i.e., species, sex, length, or age composition of catches, is a relatively straightforward problem of stratifying and randomizing the units to be sampled. Bias may develop when all market categories, types of gear, kinds of vessels, fishing grounds (within the area occupied by the stock), and landing ports are not included in the sample. It is desirable to consider each possible unit as a stratum to be sampled at random. A more subtle but still serious bias may arise from selection of the fish for measuring. Commonly, the larger fish are on top of containers because they either tend to "surface" when dumped or are placed there to attract buyers. Hence it is desirable to measure the fish in either entire containers or portions from top to bottom. Another source of bias is related to time and handling after capture. Fish in rigor mortis are significantly shorter than they are when they are not in rigor and have been pressed beneath other fish in the hold. Oysters may become shorter after handling because of breakage of the thin edges of the shell. Fortunately, these biases may be detected fairly easily by small comparative studies and may be either avoided or measured.

Different and even less tractable problems arise in determination of a standard measure of the fishing effort. The measure of effort is affected not only by the variability in space and time of the fish stocks themselves, but also by the variability in habits and efficiencies of the fishing units that exploit them. Difficulties arise not so much because of the difficulties of obtaining samples that form a large fraction of the landings, but because of the nonrandom behavior and changes in efficiencies of fishing nets. Fisher-

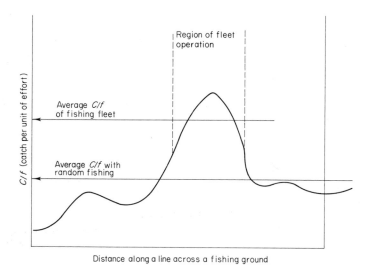

Fig. 6.10.   Diagram of the distribution of *C/f* across a fishing ground.

men tend to concentrate where they think fish are abundant; hence, usually the average catch per unit will be above the average expected from the area occupied by the whole stock (Fig. 6.10). They may lose efficiency by concentrating so much that they interfere with each other, by being unable to find concentrations and spending a large amount of time scouting, or by missing a seasonal run of fish as a result of a price dispute or weather factors. On the other hand, they may gain in efficiency by learning better how to find and catch fish through experience and by using better navigational aids or fish-finding devices. An even more important and subtle source of variability is the steady improvement in vessels and gear. Commercial fishermen try to obtain vessels that are larger and more powerful, that can handle bigger fishing gear, reduce running time, and lose less time from repairs in port or weather at sea. Such changes have brought increases in efficiency of fishing units of 50% or more within a few years.

Accounting for such variations and trends requires an intimate knowledge of both fish and fishermen. The uneven distribution of effort can be accounted for by division of the area and time into small units and suitable weighting of each. Changes in efficiency of fishing units can be accounted for by a determination of the ratios of relative fishing abilities of different gears or vessels by special comparison.

6.4 Availability and Gear Selectivity

After determining the structure of the catch from a stock by a sampling program, the biologist estimates the structure of the populations comprising the stock. The catch may be a large fraction of the stock and it may be sampled with care, yet its structure will be certain to differ from the structure of the populations because of differences in availability of the animals and the selectivity of the gear used.

Availability is defined generally as the fraction of the stock that lives in areas where it is susceptible to (i.e., may encounter) a given fishing gear during a given season. Availability varies commonly with species, sex, and size of the animals, and each of these factors may vary with time or location. This variability arises primarily from different feeding and spawning migrations and different behavior. For example, the rosefish of the Northwestern Atlantic can be caught by ordinary trawls in the day but not at night because they move up in the water; the large individuals among the halibut of the Northeastern Pacific prefer rough bottoms and the small individuals prefer smooth bottoms; the 1-year-old and 2-year-old members of the fur seals of the Bering Sea remain at sea during the breeding season, when the 3-year-old and 4-year-old males are available for capture on the shore; in many species males and females tend to segregate prior to or after spawning;

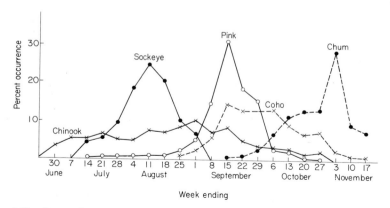

Fig. 6.11. Seasonal occurrence of salmon in Fraser River gill net catches. [From G. A. Rounsefell and G. B. Kelez (1938). The salmon and salmon fisheries of Swiftsure Bank, Puget Sound, and the Fraser River. *U.S., Fish. Wildl. Serv., Bur. Fish.* **49**, 693–823.]

and in many species size groups tend to segregate in pursuit of food. When the stock includes more than one species, the availability will differ because no two species can occupy exactly the same niche (Fig. 6.11).

The availability to different kinds of fishing gear is certain to differ also for reasons other than the selectivity of the gear (discussed next). Each combination of gear and vessel works better in certain places than in others: bottom trawl nets require relatively smooth bottom, line gear can operate in canyons and other rocky areas, purse seines can fish from the surface to a

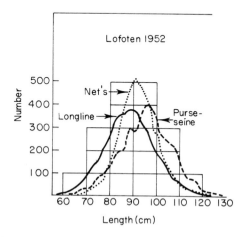

Fig. 6.12. Length composition of Norwegian cod taken with different fishing gear. [After: G. Rollefsen (1956). Introduction to problems and methods of sampling fish populations. *Rappo. Proces-Verb. Reunions, Cons. Perma. Int. Explor. Mer* **140**, 5–6.]

considerable depth, etc. For example, the introduction of purse seine gear
in the Lofoten cod fishery produced catches of fish larger than those taken
by either gill nets or long line gear (Fig. 6.12). The purse seines also produced
occasional catches of up to 90% of a single sex.

Even though individuals in a stock encounter the fishing gear they may
not be caught because the meshes or hooks are selective. Generally the
meshes of trawls and seines will retain fish whose girth is greater than approximately the circumference of the opening, but the meshes of a gill net
will hold only those fish whose girth and shape permit them to wedge in
the meshes; smaller fish wriggle through and larger fish twist out. Hooks
tend to retain only fish of a limited range of sizes because large fish are able
to break free and small fish are unable to grasp the bait. The meshes and
slats of pots used to catch crustaceans allow small individuals to escape.

The relation between size of mesh and size of fish retained depends, not
only on the relation of mesh circumference to fish girth, but also to a lesser
degree on the size of the net twine, stretch of the fiber, and hanging of the
net, i.e., the way in which the net spreads out while fishing. With some kinds
of nets and species the behavior is also important because in parts of trawls

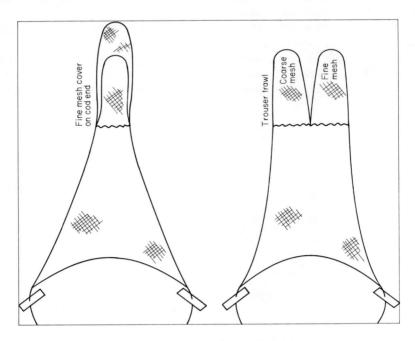

Fig. 6.13.   Diagram of otter trawl nets rigged for mesh selectivity experiments. The mesh
in the forward part guides the fish to the cod end where the filtering occurs.

and seines a large mesh with small twine will guide fish to a bag where the actual filtering occurs. Thus, commonly, the selectivity of nets must be determined by comparative fishing experiments. In these experiments nets of different construction are fished side by side and the operation is repeated as many times as are necessary to minimize the variance between the means of successive sets. One can determine the selection of trawl nets also by covering the bag with a finer mesh or by rigging a special trouserlike bag with different mesh in each leg (Fig. 6.13).

The catch by species and sex from such comparative fishing is measured, and the results are expressed as a *length selection curve*, or ogive. The percentages of each length group retained are computed and plotted (Fig. 6.14), and, when necessary, are adjusted to satisfy the assumption that the fishing powers of the nets are equal. Selection curves are compared normally at the points at which 50% of the fish are retained.

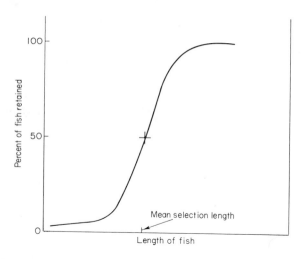

Fig. 6.14   Hypothetical selection curve for a trawl net. The lower end of the curve may not reach zero because some small fish become mixed with the bulk of the catch.

6.5   Recruitment

Recruitment, R, is defined most commonly as the number of fish of a single year group entering the exploitable phase of a stock in a given period by growth of smaller individuals. It is defined also as the number of fish from a single year group arriving in an area during a given period where fishing is in progress even though the fish may be so small that chance of capture is negligible (e.g., at the time that larval stages descend to the bottom). The latter is a more definite biological event than the former, but generally, the former is measurable and the other is not. The term shall be used with the former meaning and with the added advantage that in many stocks of fish the individuals mature at about the same time that they become exploitable; the size of the exploitable stock is an index of the number of spawners.

Recruitment is of special interest, first, because of its relation to subsequent yield of the year group. It is the first obtainable index of the abundance of the year group and is therefore a useful base for prediction. In the special instance of stocks that produce only an occasional *dominant year class* (Fig. 6.15), the appearance of abundant young is a sign of good fishing to come.

More important to understanding the dynamics of the stock is the relation between the number of spawners and the ensuing recruitment. This relationship is one of the two key relationships to be determined in the analysis of a fish stock; the other is the relation between the amount of fishing and catch for a given recruitment.

The general relation between the number of spawners and the ensuing recruitment conforms to three factors derived from the concepts of the growth of single-species populations (see Chapter 5). First, when there are no spawners there is no recruitment exclusive of immigration, which we would account for separately. Second, all populations except those headed for extinction have a capacity for growth, a resiliency that enables them to fill a niche and recover from adversity. Third, populations are limited in number by natural factors that increase the mortality rate as the number of animals increases. These factors and the ensuing mortality are called *compensatory*. Other *density-dependent* factors become less effective as density increases and are called *depensatory*.

The compensatory factors affecting fish populations include the following (after Ricker, 1954, pp. 562, 563):

1. Prevention of breeding by some members of large populations because all breeding sites are occupied. Note that territorial behaviour may restrict the number of sites to a number less than what is physically possible.

2. Limitation of *good* breeding areas, so that with denser populations more eggs and young are exposed to extremes of environmental conditions, or to predators.

3. Competition for living space among larvae or fry, so that some individuals must live in exposed situations. This too is often aggravated by territoriality—that is, the preemption of a certain amount of space by an individual, sometimes more than is needed to supply necessary food.

4. Death from starvation or indirectly from debility due to insufficient food, among the younger stages of large broods, because of severe competition for food.

5. Greater losses from predation among large broods because of slower growth caused by greater competition for food. It can be taken as a general rule that the

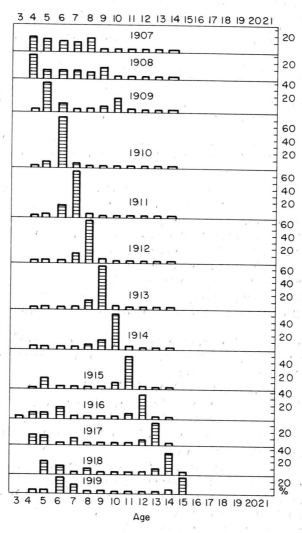

Fig. 6.15.   Percent age composition of samples of the catch of spring herring along the Norwegian coast. Note the predominance of the 1904 year class. [From J. Hjort (1926). Fluctuations in the year classes of important food fishes. *J. Cons., Cons. Perma. Int. Explor. Mer* **1**, 5–38.]

smaller an animal is, the more vulnerable it is to predators, and hence any slowing up of growth makes for greater predation losses. Since abundant year-classes of fishes have often been found to consist of smaller-than-average individuals, this may well be a very common compensatory mechanism among fishes.

6. Cannibalism: destruction of eggs or young by older individuals of the same species. This can operate in the same manner as predation by other species, but it has the additional feature that when eggs or fry are abundant the adults which produced them tend to be abundant also, so that percentage destruction of the (initially) denser broods of young automatically goes up—provided the predation situation approaches the type in which kills are made at a constant fraction of random encounters.

7. Larger broods may be more affected by macroscopic parasites or microorganisms, because of more frequent opportunity for the parasites to find hosts and complete their life cycle.

8. In limited aquatic environments there may be a "conditioning" of the medium by accumulation of waste materials that have a depressing effect upon reproduction, increasingly as population size increases.

Floods, droughts, extreme temperatures, and other environmental changes may be noncompensatory, but even these nonbiological factors may be more devastating to large populations than to small ones (and hence compensatory) by killing the individuals that cannot find shelter.

Compensatory mortality factors appear to affect most the younger stages of aquatic animals—the eggs, larvae, and juveniles—which are produced in great abundance by many aquatic animals. Such factors affect the mature individuals also, but probably much adult mortality is noncompensatory. In many stocks the mortality due to fishing is primarily among adult individuals.

When we represent the relation of recruitment to the number of spawners, it is convenient and frequently accurate to use the size of the stock as an index of the number of spawners. We expect a curve showing an average relationship within the limits shown in Fig. 6.16. All such curves must go through the origin and cross the line at which recruitment just replaces the parent stock or the "45-degree line." If they remained entirely above the 45-degree line, they would be increasing without limit; and if they remained entirely below, they would be headed for extinction. At higher levels of parent stock the recruitment may either approach an upper limit or decline to near zero. When the curve tends toward an upper limit, the population size tends toward stability; whereas when the curve descends sharply toward a zero recruitment, the population size tends to oscillate.

Stock-recruitment curves tend to be obscured by variability in the relationship. Great variability and lack of demonstrable relationship seem to be associated with species having young in large numbers and with very high mortality rates. These are the species that tend to produce occasional dominant year classes. For some such species (Fig. 6.17) there appears to be little reason for concern about a scarcity of spawners because a few of them

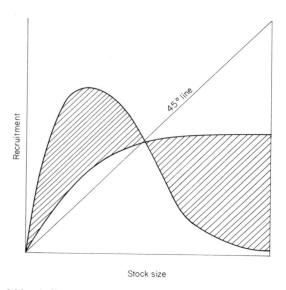

Fig. 6.16. A diagram of the bounds of average recruitment curves.

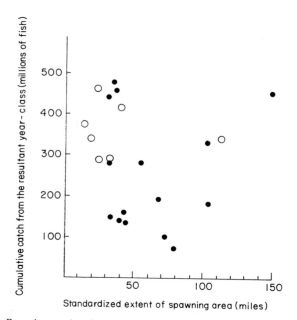

Fig. 6.17. Recruitment data for British Columbia herring. (●) West coast of Vancouver Island, 1937–1952. (○) Lower east coast of Vancouver, 1947–1953. [From N. Hanamura (1961). On the present status of the herring stocks of Canada and Southeastern Alaska. *Bull. Int. N. Pac. Fish. Comm.* **4**, 67.]

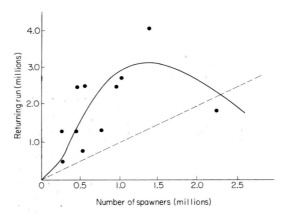

Fig. 6.18. A stock-recruitment curve for sockeye salmon in the Wood River, Alaska. Stock equals number of spawners and recruitment equals number returning from the sea. (---) Recruitment = number of spawners. [Data from F. J. Ossiander, ed. (1967). Bristol Bay red salmon forecast of run for 1967. *Alaska Dep. Fish Game Inform. Leafl.* **105**, 1–51.]

seem as likely to produce a dominant year class as a multitude. When the relationship is evident, even when there is variability about the relationship (Fig. 6.18) it may provide a basis for a superior management strategy (see Section 9.3).

6.6 Stock Size

The *stock size* is either the number, N, of individuals in, or the weight, P, of a stock at, a given time. Both measures are usually necessary, so it is customary to use information on size composition and length–weight relationship for conversion of one to the other. The terms *abundance* and *density* are used by some authors to mean the total number of individuals, but these terms are assigned more properly to either number per unit of area or unit of fishing effort. The term *biomass* may be applied to the weight of the stock. Some authors use *population* with approximately the meaning assigned to *stock* in this volume and hence use *population size*.

The stock size *per se* in terms of P is useful as a measure of the utilization of aquatic space. The total weight of one species or of all species of fish in a body of water may be reported as the *standing crop* in units of weight per unit of area as a rough index of productivity.

The change in stock size is much more useful. It may be related to corresponding changes in the environment, in fishing, in populations of food organisms, predators, and competitors. Further, the change in stock size, or in the size of some of its parts, provides the fundamental statistic for estimation of survival and mortality rates. The determination of stock size is, therefore, an important step in the study of an exploited stock. Many scientists have devised a variety of methods.

6.6.1 DIRECT COUNTS

Whenever this method can be used it is preferred because it is usually cheaper and more precise. Fortunately, it is possible in a wide variety of environments. It is usually accomplished by counting the individuals in a known fraction of the area occupied by the stock and capturing simultaneously a sample in order to add information on the size and age structure. When a pond can be drained without the loss of fish, the entire stock may be available for examination. Salmon migrating up a river can be counted from observation towers (Fig. 6.19) or booths in fishways with a high degree of accuracy and sampled occasionally with a seine. The accumulation of spawning salmon in a stream can be estimated with fair accuracy by trained observers in airplanes. Sessile animals, such as clams or oysters, can be counted or examined in *quadrats* selected at random and containing a known fraction of the beach area. Clams and other invertebrates in deeper water

Fig. 6.19. Migrating salmon are counted easily in the Kvichak River, Alaska, from a tower placed near the path of migration. Surface ripples are reduced by a small boom. (Photo by W. F. Royce, University of Washington, Seattle, Washington.)

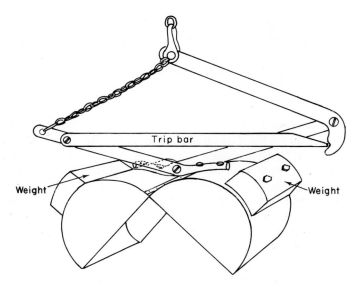

Fig. 6.20. A Petersen type dredge is used to sample animals on sandy or soft bottoms.

can be sampled with a dredge (Fig. 6.20) that is known to bring up a definite area on the bottom. Fish with territories in streams or on coral reefs can be counted by SCUBA divers.

The catch of any mobile fishing net can be used for direct estimation of stock size when the selectivity of the net and the area covered are known. Both of these matters are difficult to determine, and so this technique is rarely used. Instead the selectivity is standardized, a constant efficiency is assumed, and the catch per unit of effort, C/f, is used as an index (see Section 6.6.4).

The adaptation of sonar gear for fish detection has stimulated numerous attempts to use the instruments to count fish (Fig. 2.12). As with fishing nets, the selectivity of the echoes and the area covered must be determined. With the addition of a counter or integrator to the electronic equipment, the instrument will yield a total count in addition to a record of the echoes. Such sonic equipment must be used in conjunction with fishing nets, however, for reliable identification of the species of animals causing echoes.

6.6.2   Correlated Populations

Sometimes breeding populations of aquatic species can be estimated from the production of eggs or the numbers of nests (or redds for salmon). Estimation from eggs has been attempted for a few marine species that have pelagic eggs. The information required is the average number of eggs per female or per unit of weight of females, the sex ratio of mature animals, and the total number of eggs laid in the sea during the spawning season. The former is relatively easy to obtain, but the latter is extremely difficult to procure because of sampling problems. The entire geographic range must be sampled with plankton nets during the complete period of spawning. The period may be much longer than the time needed for hatching and larval development of a single batch of eggs. The eggs and larvae are subject to high rates of mortality; therefore the rates of development and mortality must also be known and adjustments made. Despite these difficulties, the numbers of eggs have provided important clues to the presence of large fish populations that are not being exploited, e.g., anchovies off the coast of California.

A rough estimate of the numbers of eggs laid by spawners has been developed for the stocks of herring along the coasts of British Columbia and Alaska (Fig. 6.17). The eggs are laid in the intertidal zone, where they adhere to the bottom or to vegetation. Estimates are made visually of the length of beach covered (in miles) and the density of spawn. These may be converted very roughly to an estimate of the total number of eggs by a count of the eggs in a subsample of the beach and an estimate of the number of females from data on fecundity.

The nests of certain fish, such as salmonids and basses, may be counted and the count used as an estimate of the number of spawning pairs. Counts of Pacific salmon nests (redds) are made at times to provide information on the utilization of stream areas for spawning. Usually such counts are made in limited and easily accessible areas and are considered to be an index because sampling of all spawning areas and during all periods may be difficult.

6.6.3 MARKED MEMBERS

Marked or tagged fish may provide information on stock size besides information on migrations and intermingling of stocks (see Section 6.2.2). The use of this method is indicated especially for small, discrete freshwater stocks that support recreational fisheries and for which complete catch statistics are difficult to obtain. The basic objective is to establish a population of marked fish that will be subject to the same probability of recapture as the unmarked population. Thus the fish to be marked are taken at random from catches of the gear in use and distributed at random among the unmarked population.

When a sample of fish, m, is marked and released into a stock with a number N and then another catch, C, is made, the estimated $\hat{N}$ is

$$\hat{N} = mC/r \tag{1}$$

in which r is the number of marked fish recaptured. But usually the fish to be marked must be caught and released over a period of time during which some of the marked fish may be recaptured. These data can be used in a *multiple census* (also called a *Schnabel type estimate*, after its originator), for which the simplest estimate is

$$\hat{N} = \Sigma(m_t c_t)/\Sigma r_t \qquad (2)$$

in which c_t is the total sample on day$_t$, m_t the total marked fish at large on day$_t$, and r_t the recaptures on day$_t$. An example of the computations is given in Table 6.1.

TABLE 6.1

A MULTIPLE CENSUS OF CRAPPIES FROM THE
NORTH HALF OF FOOT'S POND, INDIANA[a,b]

5–Day period	$\Sigma m_t c_t$	Σr_t	$\hat{N}$
1	2850	1	2850
2	5710	3	1900
3	8410	4	2100
4	12,470	5	2490
5	21,660	9	2410
6	36,540	15	2440
7	36,540	15	2440
8	45,980	17	2700
9	61,060	19	3210
10	67,900	22	3090

[a]Recomputed from data used by Ricker (1958).
[b]Recaptures were made in traps used to catch the fish for marking.

All of which is deceptively simple because of the difficulty of establishing a marked population of sufficient size to yield an adequate number of re-captures and subject to the same probability of recapture as the unmarked population. The pitfalls are numerous and include the following:

1. The formulae given above are positively biased approximations when numbers of recaptures are small. They have been modified by several people to provide more precise estimates.
2. The marked population may decrease in size during the experiment because of deaths caused by marking or loss of marks.
3. The unmarked population may vary in size during the experiment because of recruitment.

4. The rate of exploitation on the marked and unmarked populations may differ because the marks make the fish more liable to capture (some tags catch on nets) or less liable to capture (some marked fish are injured or disturbed and behave differently).
5. Marks may not be recognized and reported by fishermen.
6. The marked fish do not distribute themselves at random in the stock, and random sampling of the stock is not possible. Some fish may have territories near fixed gear and be recaptured repeatedly. The stock may range over bottoms and depths that cannot be fished. The range of migration of the stock may not be known accurately.

Despite these difficulties the method is the only one possible under certain circumstances, and investigators use it with special efforts during the design of the experiment and the analysis to compensate for any known problems.

6.6.4   CATCH PER UNIT OF EFFORT

Most stocks of fish in the sea and in large lakes do not behave in ways that permit either direct or correlated counts, and they are either too large or too remote to permit estimation of stock size by means of marked members. For such stocks the catch per unit of effort, C/f, is a useful index of the size. Even without a measure of the relation of the index to the actual N, frequently one can assume reasonably that each unit of a fishing operation, say one day's fishing by a trawler, captures a fraction k of the stock being fished. It follows that f units capture a fraction fk of the stock and

$$C = fkN, \; C/f = kN \tag{3}$$

or, in words that k is an unknown constant.

Much can be done to insure the assumption that k is constant. The fishing effort is expressed usually as a unit of time fished by standard units of gear and vessel, such as angler hours, or trawling hours times horsepower, or "soaking" time of fixed nets. Account is taken of nonfishing time spent in travel and lost to bad weather. Any cyclical changes in behavior of fish with either days or seasons are averaged out. Any long-term trends in either gear efficiency or skill of fishermen as a group are used to adjust the measure of effort.

6.7   Survival and Mortality

Survival, S, and mortality, $1-S$, its counterpart, are determined by a comparison of the numbers of a cohort or group of cohorts alive at successive ages. The cohort(s) may be composed either of marked animals or a year class(es) of animals in a stock as determined by age analysis and sampling of the catches.

A comparison of this section and the preceding one on stock size will reveal that the basic data for determination of stock size and survival rate of a cohort are similar if not identical in many cases. It is obvious from formula (1) that m can be considered as a cohort and

$$r/m = 1\text{-}S \qquad (4)$$

in which $1-S$ is the estimated fraction caught and S is the estimated fraction surviving if we assume no natural mortality. The similarity is so close that many authors have considered determination of stock size and of mortality rate as a single topic. It is desirable, however, to consider these methods separately because most methods of determining stock size are useful for either stocks or fractions of stocks of unknown age composition, whereas the methods of determining mortality rate for changing cohort size are useful usually for stocks with a known age structure.

The student who reviews the literature in this field will find a variety of symbols and terminology. The major recent authors, Beverton, Holt, Ricker, and Schaefer, have each tended to use different symbols. The closest approach to agreement on symbols is the list (Table 6.2) recommended by the International Council for Exploration of the Sea at its 1957 meeting.

TABLE 6.2
TERMINOLOGY AND NOTATION FOR FISHERY DYNAMICS [a]

Symbol	Definition	Term
N	Total number of fish in stock	Stock number
P	Total weight of fish in stock	Stock biomass
C	Total number of fish in catch	Catch in number
Y	Total weight of fish in catch	Catch in weight, yield
Z	Instantaneous total mortality coefficient $= -dN/Ndt = F + M$	Total mortality coefficient
F	Instantaneous coefficient of mortality caused by fishing	Fishing mortality coefficient
M	Instantaneous coefficient of mortality by (natural) causes other than fishing	Natural mortality coefficient
R	Number of fish entering the exploitable phase of a stock in a given period	Number of recruits, (annual) recruitment
l	Length of a fish	Fish length
w	Weight of a fish $w \text{ (stock)} = P/N$ $w \text{ (catch)} = Y/C$	Fish weight

TABLE 6.2 (*continued*)

Symbol	Definition	Term
t	Time, either absolute or in terms of life-span of fish	
t_r	Age at which fish are recuited a fishable stock	Age at recruitment
t_c	Age at which fish are first liable to capture by the fishing gear in use	Age at first capture
l_c	Length of fish at age t_c	Length (size) at first capture
w_c	Weight of fish at age t_c	Weight at first capture
l_r	Length of fish at age t_r	Length at recruitment
w_r	Weight of fish at age t_r	Weight at recruitment
X	Fishing effort, this symbol should be used only if no ambiguity will arise with statistical notation in this context; otherwise the following notation (g, f) is to be preferred, and used in any case when distinction is to be made between uncorrect fishing effort statistics, as recorded, and the *effective overall fishing intensity* computed from them	Fishing effort
g	Fishing effort as recorded	Uncorrected fishing effort
f	Weighted mean fishing effort per unit area, expressed in standard units and calculated with weighting factors equal to the density of fish in each area	Effective overall fishing intensity
S	$e^{-z}(-Z)$, (mean) survival rate	Fraction surviving
$1-S$	Mortality rate	Fraction dying
E	$(1-S)F/Z$, Unconditional fishing mortality rate, expectation of death by capture	Fraction caught
D	$(1-S)M/Z$, Unconditional natural mortality rate, expectation of death by natural causes	Fraction dying a natural death
q	F/f	Catchability coefficient
$Y/f\ C/f$ $Y/g,$ *or* C/g	Catch per unit effort in weight or numbers, corrected or uncorrected	Catch per unit effort

Suggested suffixes:		
i or x	Year-class	Year-class
j or n	Age-group	Age-group

[a]After S. J. Holt (1960). "Multilingual Vocabulary and Notation for Fishery Dynamics," Food and Agr. Organ. U. N., Rome.

6.7.1 FRACTIONS AND RATES

It is essential to change from the fractions S and $1-S$ to instantaneous rates for studies of mortalities. Let us assume that the rate of decrease is proportional to N in a cohort, then the rate of decrease at age t is

$$dN/dt = -ZN \qquad (5)$$

in which Z is the instantaneous total mortality coefficient.* This is related to the fraction S,

$$S = \frac{N_1}{N_0} = e^{-Zt} \qquad (6)$$

and if one should need to compute the number surviving at some time t after t_0

$$N_t = N_0 e^{-Z(t-t_0)} \qquad (7)$$

Usually computations of mortality involve both natural mortality, M, and fishing mortality, F. If we assumed that these occur simultaneously and constantly during period t, then

$$F + M = Z \qquad (8)$$

Sometimes it is useful to divide $1-S$ into the fractions dying from fishing, E, and from natural causes, D. These are related to F and M, as

$$E = \frac{(1-S)F}{Z} \quad \text{and} \quad D = \frac{(1-S)M}{Z} \qquad (9) \text{ and } (10)$$

As a simple example of such rates, let us consider again the pond with no spawning area that is stocked with 1000 catchable-sized trout and is found

---

*The concept of instantaneous rates troubles many students who may be helped by an example. Let us assume that a population of fish at the beginning of a year is $N_0 = 1000$ and subject to an annual expectation of death of 0.8. Survival, of course, is 0.2. It follows that the mortality during a year is

$$0.8 \cdot 1000 = 800, \qquad N_1 = 200$$

Now suppose that the year is divided in half and the annual expectation of death also is divided in half so that in the first half year

$$0.4 \cdot 1000 = 400, N_{\frac{1}{2}} = 600$$

In the second half of the year the same mortality rate is applied to the remaining fish.

$$0.4 \cdot 600 = 240, N_1 = 360$$

and total mortality is $400 + 240 = 640$ instead of 800.

If the year is divided into four parts similar computations indicate that $N_1 = 410$; if into ten parts, $N_1 = 431$; if into a very large number of parts, $N_1 = 449$. The latter is the result of an instantaneous rate, $Z = 0.8$, divided n times (with n very large) and applied to n fractions of the period. In this example the fraction dying in one year is $(1 - S) = 0.551$. Z is always larger than $1 - S$ and may be greater than 1.

to contain 600 when it is drained a year later. The catch resulting from 50 days of fishing effort during the year was 100 ($E = 0.1$), and the presumed deaths from natural causes were 300 ($D = 0.3$). Thus,

$$S = \frac{600}{1000} = 0.6$$

From (6):

$$S = e^{-Z(1)}$$

Changing S to natural logarithms and substituting, we have

$$log_e\, 0.6 = -Z = -0.511$$

The instantaneous total mortality rate, Z, is equal to the natural logarithm (with the sign changed) of the fraction surviving, i.e., the complement of the fraction dying. Note that this is an annual unit of the instantaneous rate. For a period of 12 months ($t - t_0 = 12$), the value of Z for a month would be $0.511/12 = 0.043$.

The total instantaneous mortality rate, Z, may be divided into M and F if we assume that these occurred in a parallel way, by the proportion of E and D:

$$F/E = M/D = Z/(1 - S) \tag{11}$$

In this example,

$$F = \frac{0.1}{0.4}(0.511) = 0.128 \quad \text{and} \quad M = \frac{0.3}{0.4}(0.511) = 0.383$$

6.7.2 EFFORT AS AN INDEX OF FISHING MORTALITY

To assume that the catch per unit of effort, C/f, is an index of stock size, N, is equivalent to assuming that the total amount of effort is an index of fishing mortality, F:

$$f = k'F \tag{12}$$

The relative constancy of k' can be insured by the same precautions used to insure the constancy of k (see Section 6.4).

6.7.3 CATCH CURVES

Now, let us suppose that we reflood the pond and leave therein the 600 fish, undamaged by any handling. Let us suppose further that fishing effort (and fishing mortality) and natural mortality continue at the same rate,

360 fish are in the pond when it is drained at the end of the second year and 216 fish at the end of the third, and the catches in the intervening periods are 60 and 36. Let us recall that C/f is an index of N and is useful for estimating survival; thus,

$$\frac{C_1/f}{C_0/f} = \frac{N_1}{N_0} = S \tag{13}$$

$$\frac{60/f}{100/f} = \frac{36/f}{60/f} = \frac{600}{1000} = \frac{360}{600} = 0.6$$

The series of catches per unit of effort, C_i/f_i, plotted against years forms a *catch curve* that is comparable to the *survivorship curve* of demographers. On a logarithmic scale, such a catch curve (Fig. 6.21) is a straight line with a slope equal to $-Z$; in this case $Z = 0.511$.

The hypothetical population of trout in the pond used in the above example was either a cohort or a set of cohorts depending on whether all of the trout were of the same age; it makes no difference in the example. There was no spawning area, hence no recruitment. Catch was counted directly and natural deaths were estimated indirectly by a difference.

Such a hypothetical situation is similar to a population of marked individuals subject to a fishery. When we assume that natural mortality and

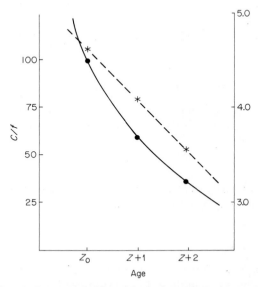

Fig. 6.21. Arithmetic (●——●) and logarithmic (*——*) catch curves from hypothetical cohort experiencing a constant mortality rate, $Z = 0.511$.

loss or nonreporting of marks occur at constant rates, then the catches per unit of effort of marked individuals provide a catch curve and an estimate of total instantaneous mortality, Z. The hypothetical situation is also similar to a cohort that is identified by determination of the age structure of a stock in successive years. When the cohort is uniformly available to the fishing gear at different ages, then the catches per unit of effort of the cohort is successive years provide a catch curve. When both recruitment and mortality can be assumed to be constant, then the age structure at any time provides a catch curve. Also, when age structures of catches from constant fishing for a series of years are available and can be averaged so that the effects of varying recruitment and mortality are smoothed, then the average age composition will provide a catch curve.

With all of these assumptions about recruitment, natural mortality, and steady states, it is surprising that so many catch curves provide useful information on total mortality. Many do resemble those in Fig. 6.22, with an ascending left-hand limb until recruitment is complete and then a nearly straight right-hand limb. The occurrence of such straight right-hand limbs is regarded as evidence that recruitment and natural mortality are sufficiently steady to give confidence in the method. Furthermore, when catch curves do not have a straight right limb, there is reason to suspect that recruitment or catchability varies or that the population is not in equilibrium.

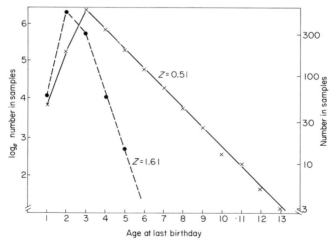

Fig. 6.22. Catch curves for the Pacific pilchard. The numbers of individuals in the samples were assumed to be proportional to the age structure of the catch. (x——x) 1925–1933. (●--●) 1937–1942. [From R. P. Silliman (1943). Studies on the Pacific pilchard or sardine (*Sardinops caerulea*). 5. A method of computing mortalities and replacements. *U.S. Dep. Interior, Spec. Sci. Rep.* **24**, 1–10.]

6.7.4 Natural and Fishing Mortality

A contrast to the relative ease and accuracy with which the total mortality rate, Z, can be determined is the difficulty of estimating the division of the total mortality rate into fishing mortality rate, F, and natural mortality rate, M, when both are occurring. Natural mortality is equal to total mortality during periods when there is no fishing but when fishing is occurring the simultaneous natural mortality must be estimated mathematically.

The natural mortality rate of catchable-sized fish in stocks that have never been fished is related to the maximum age of the fish to be expected. For example, if fish were of recruitment size at age 3 and sampling indicated only 1% as many fish at age 13 as there were at age 3, then the average natural mortality rate, $M = (\log_e 100 - \log_e 1)/10 = 0.46$. If only 1% remained after 20 years, then the average $M = 0.23$. (Note the estimated mortality rate of 0.235 for an unfinished halibut population attaining about 25 years of age as shown in Fig. 5.5.).

Where fishing ceases for a time and then resumes, the natural mortality in the interim may be especially useful. For example, an estimate of the mortality of plaice in the North Sea was obtained from measurements of the abundance of the 1932 and 1933 year classes before and after the six-year cessation of fishing during World War II. This value as given in Graham (1956, p. 391) was about $M = 0.1$.

When such special circumstances do not permit direct estimates of natural mortality, the method derives from the relationship between total mortality rate, Z, and total fishing effort, f, when both are varying. An estimate of M for the California stock of pilchard was obtained by Silliman (1943) from catch curves for two separate periods, within each of which the recruitment, R, stock size, N, and effort, X, were relatively stable but between which the effort, X, and total mortality rate, Z, were greatly different. The catch curves (Fig. 6.22) showed that $Z = 0.51$ for the period 1925–1933 and $Z = 1.61$ for the period 1937–1942, when X was four times as great as for the first period. Assuming that natural mortality, M was equal in the two periods and that fishing mortality, F, was in the same proportion as total effort, X, (availability was equal in the two periods), one may write two simple, simultaneous equations:

$$F_1 + M = 0.51$$
$$F_2 + M = 1.61$$

and

$$F_2 = 4F_1$$

The solution (rounded off) is

$$F_1 = 0.37$$
$$F_2 = 1.46$$

and

$$M = 0.15$$

A graphical solution of the same problem (Fig. 6.23) illustrates the principle involved—extrapolating mathematically to zero fishing effort.

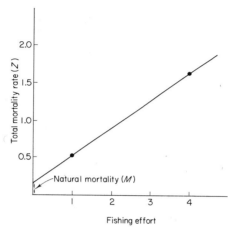

Fig. 6.23. A graphical method of estimating the natural mortality rate for the Pacific pilchard.

If fishing effort, X, and total mortality should both vary continuously, then their relationship would provide a similar possibility of estimating M. An illustration in the form of a simple linear regression is shown for Georges Bank haddock in Fig. 6.24. In this example, $M \simeq 0.0$ although this value is not very reliable because the variability in the data is relatively large. In addition, a mathematical refinement is desirable to overcome the slight error of using the average of a year's data to estimate the abundance at the midpoint of the year.

Regardless of the difficulties in determining M when fishing is occurring, most estimates are lower than the M that would be expected on the basis of maximum age of the animals in unfished populations. M is frequently estimated to be only about 0.1 or 0.2 under conditions of heavy to moderate fishing on stocks that fall far short of reaching the 25–50-year ages that would

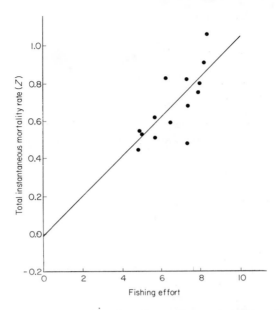

Fig. 6.24. Extrapolation of relation of total mortality rate to fishing effort by linear regression analysis for estimation of natural mortality rate. [Data for Georges Bank haddock from 1933–1945 from H. A. Schuck (1949). Relationship of catch to changes in population size of New England haddock. *Biometrics* **5**, 213–231; and W. F. Royce and H. A. Schuck (1954). Studies of Georges Bank haddock. Part II. Prediction of the catch. *U.S., Fish. Wildl. Serv. Fish. Bull.* **56**, 1–6.]

occur in unfished populations subject to these natural mortalities. It is clear that the reduction in stock size that accompanies fishing diminishes the natural mortality rate—a compensatory change from reduced environmental pressure on the survivors.

6.8 Growth

Growth of aquatic animals is a highly irregular process (see Section 4.7); it varies with age, sex, season, climate, reproductive cycle, and population size. It is especially irregular during larval and postlarval periods, as rapid changes take place in body form. A complete description of growth in mathematical terms has not yet been formulated although many attempts have been made.

One regularity is present, however, in the growth pattern of many aquatic animals after the juvenile stages; their growth, in either length or weight, slows and approaches an asymptote (Fig. 6.25). When the average length

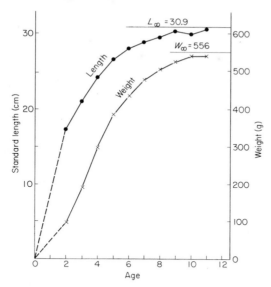

Fig. 6.25. Average lengths and weights of ciscos ranging in age from 0–12 years from Vermilion Lake, Minnesota. [Data used by Ricker (1958), from K. D. Carlander (1950). "Handbook of Freshwater Fishery Biology." Brown, Dubuque, Iowa.]

of individuals in a year group, l_t, is plotted against the average length of a group one year older, l_{t+1}, the approach to zero growth is clearly shown (Fig. 6.26).

For many species of fish this general growth pattern is approximately

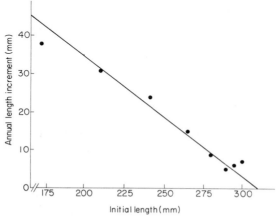

Fig. 6.26. Graphical method of estimating L, illustrated with data used in Fig. 6.25.

exponential and is adequately expressed by an equation ascribed to von Bertalanffy,

$$l_t = L_\infty \left(1 - e^{-k(t-t_0)}\right) \tag{14}$$

which can be incorporated readily into yield models. It has a simple transformation to weight measurements,

$$w_t = W_\infty \left(1 - e^{-k(t-t_0)}\right)^3 \tag{15}$$

but a simple cubic function of weight to length is assumed. Note that the length curve is negatively exponential throughout its length, whereas the weight curve is positively exponential in early life and later negatively (see Fig. 4.21).

Methods of fitting these curves are given by Ricker (1958), Gulland (1969), and others. When the data are not fitted adequately by these curves, functions found in Richards (1959), or in Parker and Larkin (1959) may be used.

6.9 Yield Models

Let us now bring together all of the factors that have been described earlier as influencing the size of a stock and develop models that will describe the effects of past fishing and predict the effects of changes in the fishing. Let us assume that we have identified a unit stock subject to fishing and try to estimate the average (called also the *equilibrium* or *sustainable*) yields under various amounts of fishing. The unit stock and the factors affecting its size may be diagrammed as in Fig. 6.27.

Fig. 6.27.   Diagram of the additions to and losses from a stock.

All models are based on this general relationship. In an unfished stock the size is presumed to be at an equilibrium level as a result of the natural controlling factors. As fishing mortality increases, stock size diminishes, natural mortality reduces, growth increases, and recruitment may or may not change for a time. If fishing mortality should increase sufficiently, however, recruitment would diminish. It is assumed that a level or levels of fishing mortality can be found that permit a maximum sustainable yield.

6.9.1 THE LOGISTIC MODEL

The most generally applicable model is the relation of stock size to fishing mortality rate. Stock size is usually measured by the catch per unit of effort; fishing mortality rate by carefully refined measures of the total effort. Application of the model requires a long series of data and no attempt is made to identify separately the rates of recruitment, growth, and natural mortality. Any changes in these factors caused indirectly by fishing are assumed to have reached equilibrium.

The model derives from the prevailing tendency for population growth to follow a sigmoid curve (see Section 5.15) in which the rate of natural increase is at a maximum at the point of inflexion and decreases gradually to zero either at zero population size or at the limiting population size. In an equilibrium state the catch is equal to the rate of natural increase and the stock size is a function of fishing effort. Data from some fisheries indicate that a simple linear relationship is adequate, thus

$$dP/dt = k_1 P(L - P) \tag{16}$$

in which P is the total weight of fish in the stock, L the limiting weight of the stock and k_1 a constant. The equilibrium yield in pounds, Y_e, may be represented adequately by

$$Y_e = k_2 f P \tag{17}$$

in which k_2 is another constant and f the fishing effort. Since

$$Y_e = dP/dt \tag{18}$$

under the assumed equilibrium conditions, and hence

$$Y_e = k_1 P(L - P) \tag{19}$$

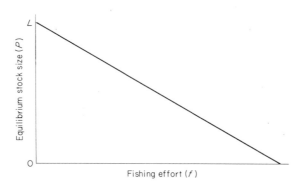

Fig. 6.28. Relation of stock size to fishing effort.

combining (17) and (19), we have

$$P = L - (k_2/k_1)f \tag{20}$$

This simple linear relation is shown in Fig. 6.28. Combining (17) and (20), we have

$$Y_e = k_2 f[L - (k_2/k_1)f \tag{21}$$

In this relationship the equilibrium yield is a simple parabolic function of fishing intensity (Fig. 6.29). Yield increases with fishing intensity until stock size diminishes to half of the maximum equilibrium size with no fishing. Further increases in fishing intensity will produce no increase in equilibrium yield but rather a decrease and a condition of overfishing.

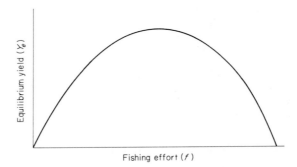

Fig. 6.29.   Relation of yield to fishing effort.

Analysis of the yield effort relationship in a number of fisheries suggests that commonly the maximum equilibrium yield is obtained at a stock size somewhat less than half the maximum—perhaps a third—and that (16) above should be a more complicated function. Nonetheless, the data available are commonly so variable that the simple equations above provide a satisfactory model.

Application of this model to actual fishery data requires the estimation on a year-to-year basis of the equilibrium yield, $Y_{e.i}$, or of the equilibrium catch per unit of effort, $Y_{e.i}/f_i$, not the actual yields, Y_i, which may be more or less than the equilibrium levels. The actual yield, Y, in each year is divided by its fishing mortality, F, for estimation of the mean population during the year:

$$Y_{e.i}/F_i \tag{22}$$

The level of population at the end of the year may be estimated as the average of the population during the year and that during the following year:

$$\tfrac{1}{2}\left(\frac{Y_{e.1}}{F_1} + \frac{Y_{e.2}}{F_2}\right) \tag{23}$$

A series of such estimates provides the data for estimating the increases or decreases during the years, that is,

$$Y_{e.i} = Y_i + (\Delta P_i/\Delta t) \tag{24}$$

The value of the fishing mortality, F, must be estimated independently either from tagging data or from a long series of data on catch and effort.

A graphical illustration of this model may be obtained from a plot of Y_e/f as an index of stock size against total fishing effort, f. When these factors are used for the coordinates, the total yield, Y, can be shown as a series of hyperbolae derived directly from the values of the coordinates. Fig. 6.30a shows a plot of the data for yellowfin tuna in the eastern tropical Pacific. In this figure the observations for successive years are connected to show time trends in the relationship. It appears that from 1943 to 1946 the fishing intensity was lower and from 1947 to 1951 the equilibrium catch per day's fishing was generally higher for the total effort than during the years 1953–1959. Nevertheless, the points fall reasonably along the estimated equilibrium line with a maximum near a total effort of 36,000 standard days (Fig. 6.30b). This relationship has formed the basis for regulatory action (although recent data indicates some changes in the effort yield relationship).

6.9.2 Dynamic Pool Model

Another general model is used when recruitment can be assumed to be independent of stock size within the range of stock sizes found and when growth and natural mortality can be measured. This one requires extensive data on catch and effort as well as data on the age and size composition of the catch. This model is much more flexible; it permits estimates of yield to be made both for different amounts of fishing effort and different ages at recruitment.

When the assumptions are made that recruitment is independent of stock size and the stock is in equilibrium, it follows that the average yield from the stock over a period is proportional to the average recruitment. It follows also that the yield from an average cohort during its life is equal to the average yield of all cohorts during any year and that the yield from a cohort is proportional to the number recruited to it. Consequently, we may use the yield per recruit without attempting to estimate the actual number recruited and expect results proportional to the yield from the stock.

As is noted above (formula 7,), the number of animals in a cohort subject

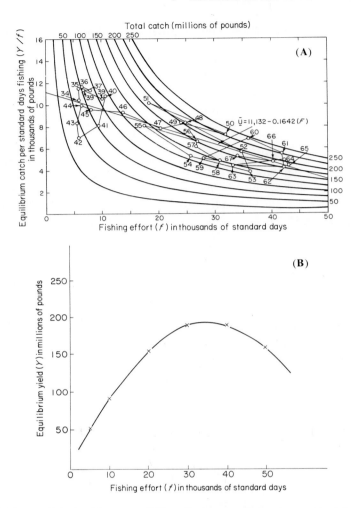

Fig. 6.30. (A) Relationships among fishing effort, apparent abundance and catch of yellow-fin tuna in the Eastern Pacific Ocean, 1934–1967. **(B)** Equilibrium yield for different amounts of fishing effort for yellowfin tuna of the Eastern Pacific Ocean. [Source: Inter-American Tropical Tuna Commission, Annual Report for 1967.]

to fishing and natural mortality after recruitment is

$$N_t = Re^{-(F + M)(t - t_r)} \tag{25}$$

The number of animals caught in the time interval $t_1 t + dt$ will be

$$C_t = F\,N_t\mathrm{dt} \tag{26}$$

so the catch, C, in the time between t_r and t_∞ (the maximum age attained—not really infinite) may be expressed by the integral

$$C = \int_{t_r}^{t_\infty} FN_t dt \tag{27}$$

and

$$C = \int_{t_r}^{t_\infty} FRe^{-(F+M)(t-t_r)}dt \tag{28}$$

Integrated, we have

$$C = R\frac{F}{F+M}\left[1 - e^{-(F+M)(t_\infty - t_r)}\right] \tag{29}$$

which, when t_∞ is infinite, reduces to

$$C = \frac{FR}{F+M} \tag{30}$$

or, in words, the catch in number of animals equals the recruitment multiplied by the proportion of fishing mortality to total mortality.

We must also consider the average growth of individual animals in the formulation of the model, and to do so we follow an argument parallel to that above. The weight caught in a time interval, $t, t + dt$, is given by

$$Y_t = F_t N_t W_t dt \tag{31}$$

in which W_t is the average weight of an individual at age t. The most convenient equation for W_t is the Von Bertalanffy growth equation (see Section 6.8):

$$W_t = W_\infty \left[1 - e^{-k(t-t_0)}\right]^3 \tag{32}$$

Expanding, we have

$$W_t = W_\infty \left[1 - 3e^{-k(t-t_0)} + 3e^{-2k(t-t_0)} - e^{-3k(t-t_0)}\right] \tag{33}$$

which may be written as a summation thus:

$$W_t = W_\infty \sum_{N=0}^{3} U_n e^{-nK(t-t_0)} \tag{34}$$

in which $U_0 = 1$, $U_1 = -3$, $U_2 = 3$, $U_3 = -1$.

Integrating and rearranging, we have

$$Y = FRW_\infty \sum_0^3 \frac{U_n}{F+M+nk} e^{-nK(t_c-t_0)} \left[1-e^{-(F+M+nK)(t_1-t_c)}\right] \quad (35)$$

When $t_1 = \infty$, the last term disappears; when t_1 is large, the last term is negligible

This equation can be solved with a desk calculator (see Gulland, 1969, from which the above argument is adapted) or with the aid of tables of the Incomplete Beta Function (Wilimovsky and Wicklund, 1963), or by means of a computer program (Paulik and Gales, 1964).

Alternatively, we may refine the arithmetical approach illustrated previously by the simple case of the trout in the pond (see Section 6.7.3). Consider the example (Table 6.3) from Ricker, 1958, of a cohort of bluegill sunfish for which the following information was determined or was assumed:

Total weight of cohort at age 2	$P_2 = 1,000$ kg
Fishing mortality (continuous throughout year)	$F = 0.5$
Natural mortality (continuous throughout year)	$M = 0.6$
Age at recruitment	$t_r = 2\frac{1}{2}$ in part
	$2\frac{3}{4}$ full
Growth rate (assumed to be exponential for quarter-year periods)	$G = \log_e W_{t+1} - \log_e W_t$
Mean length	$l_t = $ average determined for quarter-year periods
Mean weight	$W_t = $ average determined for quarter-year periods

The steps in the computations of Table 6.3 are the following as indicated by column numbers:

2 This is the mean length as determined by stock sampling
3 This is the mean weight as estimated from mean length and length–weight relationship
4 These are the differences of $\log_e$ of mean weight
5 The natural mortality rate is divided equally among the quarters of the year
6 The fishing mortality is assumed to be constant after the age at recruitment, which is approximately $2\frac{5}{8}$, and divided equally among the quarters of the year
7 This is the sum of the growth and mortality factors
8 The weight change factor is $e^{G-M/4-F/4}$, or a change of the rate of column 7 to a fractional change obtained from a table of natural logarithms

TABLE 6.3
COMPUTATION OF YIELD FOR A STOCK OF BLUEGILLS IN A LAKE[a,b]

1 Age	2 Mean length (mm)	3 Mean weight (g)	4 G	5 M/4	6 F/4	7 G − (M/4 + F/4)	8 Weight change factor	9 Weight of stock (kg)	10 Average weight (kg)	11 Yield (kg)
2	95	13	0.81	0.15	0	+0.660	1.935	1000	—	—
$2\frac{1}{4}$	109	29	0.41	0.15	0	+0.260	1.297	1935	—	—
$2\frac{1}{2}$	122	44	0.28	0.15	0.055	+0.075	1.078	2510	2608	143
$2\frac{3}{4}$	135	58	0.17	0.15	0.125	−0.105	0.901	2705	2752	321
3	145	69	0.15	0.15	0.125	−0.125	0.883	2438	2294	287
$3\frac{1}{4}$	153	80	0.13	0.15	0.125	−0.145	0.865	2152	2007	251
$3\frac{1}{2}$	160	91	0.11	0.15	0.125	−0.165	0.848	1862	1720	215
$3\frac{3}{4}$	165	101	0.08	0.15	0.125	−0.195	0.823	1579	1439	180
4	170	110	0.07	0.15	0.125	−0.205	0.815	1299	1179	147
$4\frac{1}{4}$	175	118	0.07	0.15	0.125	−0.205	0.815	1059	961	120
$4\frac{1}{2}$	178	125	0.05	0.15	0.125	−0.225	0.798	863	776	97
$4\frac{3}{4}$	182	131	0.04	0.15	0.125	−0.235	0.790	689	616	77
5	185	137	0.04	0.18	0.125	−0.265	0.767	544	480	60
$5\frac{1}{4}$	188	143	0.03	0.34	0.125	−0.435	0.647	417	344	43
$5\frac{1}{2}$	191	148	0.04	0.50	0.125	−0.585	0.557	270	210	26
$5\frac{3}{4}$	193	153	0.03	0.70	0.125	−0.795	0.452	150	109	14
6	195	158	—	—	—	—	—	68	—	—
Total			—	—	—	—	—		—	1981

[a]See Section 6.9.2.
[b]After Ricker (1958).

9 This column is obtained by multiplication of the assumed initial weight by the successive weight factors of column 8

10 This is the average of adjacent values in column 9

11 This is the product of columns 10 and 6

The total yield of the cohort is estimated as 1981 kg under these conditions.

6.9.3 YIELD PER RECRUIT WITH CHANGING RECRUITMENT SIZE

The foregoing model of the yield per recruit depends on the assumptions that recruitment is constant and is independent of stock size. It must be recalled, however, that recruitment to the catch depends on the selectivity of the gear and that in many fisheries there may be control over the minimum size of animals retained by the gear. This may be done in trawl fisheries by regulation of minimum mesh size and in hook fisheries by regulation of the minimum size of animals to be retained. Thus the yield depends on the combination of fishing intensity and gear selectivity. When the yield is maximized from the combined values, the line has been called the *eumetric fishing curve*.

The computations for determining the yield with various combinations of fishing intensity and mortality require either repeated solutions of formula (35) or repeated summations in the form of Table 6.3 with varying values of F and t_r. Ricker gives the results of such variations for the bluegill stock of Table 6.3 as follows in Table 6.4. He emphasizes that the changes in yield that are expected from changes in the selection size are rather small and notes that this characteristic seems to be typical of many fisheries. It follows

TABLE 6.4

RELATIVE YIELDS FROM THE STOCK OF
BLUEGILLS OF TABLE 6.3 WITH VARIOUS
FISHING RATES AND MINIMUM SIZES[a]

Minimum fork length (mm)	Rate of fishing (F)		
	0.3	0.5	1.0
102	76	96	110
116	77	99	120
122	76	100	125
128	75	99	128
140	71	95	125
149	65	88	119

[a]The summation of Table 6.3 is taken as 100.

that determination of an exact optimum minimum size may not be important and that moderate changes in minimum size by regulation may not cause important changes in yield.

6.9.4 SPAWNER–RECRUIT MODEL

This model is especially useful for stocks of Pacific salmon in which recruitment depends to a greater degree on the size of the parent stock and occurs mostly just before spawning. Essential also for the application of this model is the assumption that the factors limiting the size of the stock operate principally in the freshwater spawning and nursery areas and not on the ocean feeding grounds of either juveniles or adults. In this situation the model is a spawner–recruit relation of the kind shown in Fig. 6.31. It may be determined empirically from a series of data on number of spawners and ensuing recruitment. In the absence of such data for a particular stock, it may be estimated from the area of suitable spawning and nursery grounds on the basis of the production of recruits per unit of area by other stocks on other similar grounds.

6.9.5 MODIFIED YIELD MODELS

The foregoing yield models are all restricted by assumptions and comprise rather elementary models of actual fishery situations. An equilibrium state is assumed although its occurrence is not actually known, but when it is

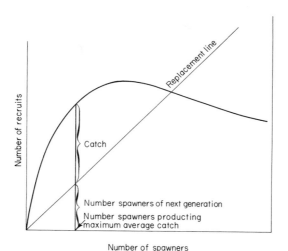

Fig. 6.31. A hypothetical spawner–recruit curve for Pacific salmon.

approximated there can be no doubt of the usefulness of the models. The decision about such an assumption is a relatively subjective one. Another prevailing assumption—not explicitly stated above—is that the niche occupied by the stock is exactly known. This is commonly not the case because of continuing exploration by fishermen or because of changes in the niche related to climatic changes. Add to these uncertainties the matters of annual cycles or irregularities in growth, natural mortality or fishing mortality, and the trends in amount and efficiency of fishing or the difficulties of measurement and the solution becomes more complicated. Such problems have stimulated the development of many modifications in the models and the application of computer simulation to deal with the large numbers of variables.

REFERENCES

The analyst of the population dynamics of exploited fishery populations will find that most of the important literature has been published since 1950 although the first significant contribution was in 1918 (by F. I. Baranov). Other important contributions were by Michael Graham, E. S. Russell, and W. F. Thompson. The major strides either in the development or in the synthesis of the science were made by W. E. Ricker, the team of R. J. H. Beverton and Sidney J. Holt, and by M. B. Schaefer during the 1950's. Many others have added important papers recently, including D. G. Chapman, L. M. Dickie, John A. Gulland, J. E. Paloheimo, G. J. Paulik, and D. S. Robson. References to these and many other authors will be found in the reference lists of the following publications.

The major general syntheses of fishery population dynamics are:

Beverton, R. J. H., and Holt, S. J. (1957). On the dynamics of exploited fish populations. *Fish. Invest. (London)* **19**, 1–533.
Graham, M., ed. (1956). "Sea Fisheries, their Investigation in the United Kingdom." Arnold, London.
Ricker, W. E. (1954). Stock and recruitment. *J. Fish. Res. Bd. Can.* **11**, 559–623.
Ricker, W. E. (1958). Handbook of computations for biological statistics of fish populations. *Bull., Fish. Res. Bd. Can.* **119**, 1–300.
Schaefer, M. B. (1954). Some aspects of the dynamics of populations important to the management of the commercial marine fisheries. *Bull. Inter-Amer. Trop. Tuna Comm.* **1**, 26–56.
Schaefer, M. B., and Beverton, R. J. H. (1963). Fishery dynamics—their analysis and interpretation. *In* "The Sea" (M. N. Hill, ed.), Vol. 2, pp. 464–483. Wiley, New York.

The following are especially valuable general texts or compilations:

Calhoun, A. ed. (1966). "Inland Fisheries Management." Resources Agency, Dep. Fish and Game, Sacramento, California.
Cormack, R. M. (1968). The statistics of capture-recapture methods. *In* "Oceanography and Marine Biology, an Annual Review" (H. Barnes, ed.) Vol. 6, pp. 455–506. Allen & Unwin, London.
Cushing, D. H. (1968). "Fisheries Biology. A Study in Population Dynamics." Univ. of Wisconsin Press, Madison.

De Ligny, W. (1969). Serological and biological studies on fish populations. *In* "Oceanography and Marine Biology, an Annual Review" (H. Barnes, ed.), Vol. 7, pp. 411–513. Allen & Unwin, London.

Gulland, J. A. (1969). Manual of methods for fish stock assessment. Part 1. Fish population analysis. *FAO Man. Fish. Sci.* No. 4, pp. 1–154.

International Council for Exploration of the Sea. (1956). Problems and methods of sampling fish populations. *Rapp. Proces-Verb. Reunions, Cons. Perma. Int. Explor. Mer.* **140**, 1–111.

International Council for Exploration of the Sea. (1964). On the measurement of abundance of fish stocks. *Rapp. Proces-Verb. Reunions, Cons. Perma. Int. Explor. Mer.* **155**, 1–223.

Nikolskii, G. V. (1969). Theory of Fish Population Dynamics as the Biological Background for Rational Exploitation and Management of Fishery Resources" R. Jones, ed., transl. by J. E. S. Bradley, Oliver & Boyd, Edinburgh.

Paulik, G. J., and Bayliff, W. H. (1967). A generalized computer program for the Ricker model of equilibrium yield per recruitment. *J. Fish. Res. Bd. Can.* **24**, 249–259.

Paulik, G. J., and Gales, L. E. (1964). Allometric growth and the Beverton and Holt yield equation. *Trans. Amer. Fish. Soc.* **93**, 369–381.

Richards, F. J. (1959). A flexible growth function for experimental use. *J. Exp. Bot.* **10**, 290–300.

Ricker, W. E., ed. (1968). "Methods for Assessment of Fish Production in Fresh Waters" Int. Biol. Progr. Handb. No. 3. Blackwell, Oxford.

Rounsefell, G. A., and Everhart, W. H. (1953). "Fishery Science, its Methods and Applications." Wiley, New York.

Rounsefell, G. A., and Kask, J. L. (1943). How to mark fish. *Trans. Amer. Fish. Soc.* **73**, 320–363.

Silliman, R. P. (1943). Studies on the Pacific pilchard or sardine (*Sardinops caerulea*). 5. A method of computing mortalities and replacements. *U.S., Dep. Interior, Spec. Sci. Rep.* **24**, 1–10.

Watt, K. E. F. (1968). "Ecology and Resource Management." McGraw-Hill, New York.

Wilimovsky, N. J., and Wicklund, E. C. (1963). "Tables of the Incomplete Beta Function for the Calculation of Fish Population Yield." Institute of Fisheries, University of British Columbia, Vancouver, Canada.

Aquacultural Science

Aquaculture, like agriculture, is an ancient art that has been advanced by scientific practices principally during the last 100–150 years. There have not been in aquaculture, however, revolutionary changes as have occurred in agriculture through the application of scientific findings to management of soils, selective breeding, nutrition, and control of disease, which have allowed a tenfold increase in the productivity of labor. Nevertheless, the scientific inputs to aquaculture are increasing rapidly, and if the changes in agriculture can be used as a precedent, similar increases in productivity can be expected in aquaculture.

Aquaculture, like agriculture, is basically a circumvention of the factors that limit natural populations. The environmental factors are controlled by choosing plants or animals that tolerate best the conditions of confinement and by giving special care to the young during their more tender ages. Food or nutrients are supplied as needed and restricted to the plants or animals in confinement. Predators and competitors are excluded. Diseases are controlled in a variety of ways.

Thus aquaculture poses challenges to many scientific and engineering disciplines. The choice of plants and animals may challenge the geneticist. The confinement, transport, or protection of animals may challenge engineers from several specialties. The care of young will challenge the microbiologist and the ecologist. The nutrition will challenge the biochemist and the physiologist. The control of disease will challenge the pathologist and the epidemiologist. Such diverse disciplines cannot be explained in this brief chapter, but their contribution to the common goal of increasing organic production from the waters shall be explained.

7.1 Control of Environment

Application of any breeding, nutritional, or disease control requires some degree of physical control. In agriculture, control may range from minimal herding of semiwild stock to intensive culture in which each individual animal or plant is bred, fed, given protection, and harvested under conditions controlled by man. Likewise, aquacultural control ranges from minor enhancement of environment to intensive culture.

The physical control required by cultivated aquatic biota depends on whether they are free during most of their lives, like fish, or sessile during most of their lives, like oysters or rooted plants. The fish require an enclosure which is usually an isolated pond, whereas the sessile animals or plants do not require a separate body of water or even barriers in the water. With either kind of organism the major objectives of control are to insure a desirable physical environment, to arrange the optimum density of the cultivated organism and to exclude all competitors or predators.

Legal control is as important as physical control with either kind of organism. Commercial aquaculture can be practiced only when those people who culture the animals or plants can also control the harvest and sale. The aquaculturist cannot possibly operate like the fisherman, who seeks wild fish and owns them only after he has caught them.

7.1.1 FISH PONDS

A pond with barriers against fish migrations at inlet and outlet and with an arrangement to allow complete draining provides control over the kinds and numbers of organisms in the pond. The animals in the pond, including fish disease organisms, can be either reduced or eliminated by either drying or chemical sterilization. The pond can then be filled with water and stocked with the numbers and kinds of animals desired. Usually the water is admitted without sterilization. It brings with it various innocuous beneficial organisms, such as phytoplankton and zooplankton, as well as a variety of pests, such as insects. Terrestrial pests which eat fish, such as birds or mammals, can be screened out if necessary. The result, which is attained in many well-managed ponds, is a population of a single species of fish under control with respect to reproduction, diet, diseases, and pests.

The control must extend, however, beyond the organisms associated with the fish to the aquatic environment. The water temperature and dissolved gases, especially oxygen, must be appropriate at all times for the species of fish. The pH must remain somewhere between 5 and 10 and preferably between 6.5 and 8.0. The dissolved salts must include the minerals essential

Fig. 7.1.   Experimental trout and salmon ponds at the University of Washington.

for the fish and not include an undue amount of either the fish's waste materials or the products of decaying organic materials.

Such a regime is used for intensively managed ponds to which all food is supplied, e.g., trout ponds (Fig. 7.1). When the water is supplied from springs in which no fish live, diseases are largely eliminated. When the temperature is near the maximum preferred by trout the year around, it allows a maximum rate of growth when the food is adequate and suitable.

On the other hand, when the pond is managed to produce all or a large part of the fish's food, as in the case of carp or milkfish ponds, the control is much more complicated. The needs of the fish during larval, growing, and breeding stages must be met and, in addition, the fish food organisms must be controlled so that food of the needed size is provided in the correct amounts at the proper time. Providing such food may involve additions of fertilizers to make available needed elements, control of the pond bottom to provide suitable spawning areas, and control of rooted vegetation to furnish the proper amount and kind. The objective is an ecological system producing all or a large part of its organic material through photosynthesis and transforming the synthesized material into flesh of a desired species as efficiently as possible.

The pond need not be managed for a single species or even for aquatic animals alone. In many ponds of Southeast Asia that produce naturally much of the animal food, two or more species that occupy different ecological niches can be grown together in order to achieve a greater total production. These can be different fish or fish and shrimp. Further, ponds can be managed for simultaneous or alternate crops of fish and rice.

7.1.2 ENCLOSURES OTHER THAN PONDS

No other enclosure for aquatic animals can be as satisfactory as a pond, but certain screens and barriers have limited usefulness. Net enclosures can hold juvenile and adult animals where they can feed on natural foods brought in by the currents as well as on artificial foods. Such enclosures are also useful to hold live animals (e.g., lobsters) for market. A major disadvantage of any aquatic net enclosure is the speed with which it becomes fouled, especially when the openings are small enough to retain larval fish. For this reason other kinds of barriers are eagerly sought, such as sound, electrical fields, curtains of bubbles; but most such substitutes have been found to be of limited usefulness. One promising barrier to limit the movement of oyster predators on the bottom is a line of sand treated with an insoluble poison.

7.1.3 BEDS AND RACKS FOR SESSILE ORGANISMS

Some larval mollusks and leafy algae attach themselves to shells or other objects. The objects, called "collectors," are arranged to suit the needs of the organisms. Some oyster, mussel, or algal collectors are strung on wires and hung from racks (Fig. 7.2) in order to raise the organisms off the bottom and out of reach of many of their pests. Such a system also allows a large number of organisms to grow per unit of area. For example, mussels may produce as much as 300,000 kg/ha/year when suspended from racks in a good growing area.

Mollusks that do not attach themselves when in the larval stage, such as clams, are commonly grown on a suitable bottom, called a "bed." Oysters can be grown in this manner also, either attached to the collectors or separated from them. Animals in a bed can be given limited protection from pests; they can be easily recaptured either for transfer to a bed with better growing conditions or for market.

Fig. 7.2. Racks from which strings of oysters are grown in Ago Bay, Kakiojima, Japan. (Photo by Arne Suomela.)

7.1.4 PARTIAL CONTROL OF OPEN WATERS

Almost every body of water is capable of producing fish, whether it is being stored, transported, or used for other purposes such as transportation or hydroelectric power. Much effort and ingenuity by biologists and engineers is directed toward maximizing the production of fish for food or recreation from waters being used for other purposes. Usually such waters are public, but are capable of being placed under some of the environmental or population controls exercised in private aquaculture. Such a program of management in open waters may be called *extensive aquaculture* in contrast to intensive aquaculture, in which the animals or plants are confined and controlled completely.

When the body of water has been modified without regard to fish production, the controls that are economically feasible may be very limited. When the water is polluted, the pollutants must be treated or diverted. When the stream is laden with silt from eroding land, the forestry or agricultural practices must be changed. When a dam has been built without a fish transport system, one might be needed. Making the necessary changes in existing practices or structures may be very expensive. It is much more efficient to plan to include fish production among the water uses before structures are built or management plans are fixed.

In ponds or small lakes the populations of organisms can be controlled to a substantial extent. Unwanted animals can be removed entirely by poisoning or partly by selective netting or spot poisoning. Unwanted plants can be cut or poisoned. Large populations of stunted fish can be reduced by destroying spawn either by changing water levels or poisoning nests. Desirable species can be introduced from hatchery stocks.

Some control can also be exercised over the water quality. Temperature below dams can be changed by varying the depth in the lake above from which the water is taken. Dissolved oxygen concentrations can be increased either by reducing organic content or by draining off the layer of water with low oxygen content above a dam or by destratifying the water. The last can be accomplished by pumping compressed air to the bottom and releasing it through small holes over a large area. Fertility can be controlled by restraining or adding sewage or artificial fertilizers. Shelter for small fish can be provided by leaving uncleared areas as land is flooded or by adding brush piles and rocks.

Streams can be managed too for enhancement of their productivity. Migratory species, especially salmon and eels, must have easy and safe passage. Natural obstructions can be removed, dams can be built with fishways or other transport devices, turbine and canal intakes can be screened. Small streams with relatively stable flow can be provided with dams and deflectors that will maintain pools and sheltered areas. When the streams are full of silt, they may benefit from the management of the farms or forests above in ways that will reduce the soil erosion.

Even the conditions for animals in the sea can be enhanced to a limited extent. Shelter can be provided by piles of rocks or wrecked autos. Grounds for rubbish dumps can be managed in ways that will attract fish.

7.2 Amenable Species

Many thousands of species of aquatic animals and plants can be used as food, but only a few species of aquatic animals provide most of the food produced by aquaculture. They are the carps, especially *Cyprinus carpio*, the true carp, the salmons and trouts, especially *Salmo gairdneri*, the rainbow trout, and the oysters of the genera *Ostrea* and *Crassostrea*. Less important species are the eel *Anguilla* in Japan and the milkfish *Chanos* in Southeast Asia. Other groups show promise of an increasing contribution: *Tilapia*, the cichlid, some catfishes, shrimp, and mollusks other than oysters. Aquatic plant culture is largely confined to a few species of marine algae in Japan and seems unlikely to expand rapidly.

In theory, almost any animal or plant can be grown in captivity. The environment can be modified to accommodate the needs of the species and the species can be acclimatized or adapted to its environment. But modifying the environment may be very costly and modifying the species may take many generations. Selecting an appropriate species will greatly cheapen and shorten the process.

The primary requirement of any food species is that it support a profitable operation. It should reproduce easily in captivity and supply an abundance of young. Its young should be hardy and easy to feed. It must be economical to feed either because it eats cheap plant materials or economically converts expensive protein food into flesh. It must also tolerate crowding in a limited space. It must either be adapted or adaptable to the water available, whether fresh or salty, warm or cold, polluted or clean, because the water cannot readily be changed. The search for amenable species or varieties is, therefore, fundamentally a search for animals to fit available aquatic environments that are not being fully utilized. The possibilities include marshes, irrigation systems, agricultural lands and waters, and waters receiving either organic waste or heat. An especially attractive possibility is the use of drainage from irrigated lands that is too salty for reuse on the land.

Aquaculture that supports angling for recreation is less restricted by cost factors and can give much greater weight to either the fighting ability or appearance of the fish than to the amount of protein produced for food. New species can be introduced into new environments in which they can be expected soon to maintain themselves or in which they must be maintained by continued stocking. Exotic species such as trout in tropical mountain streams can be raised so that a great variety of recreational opportunity can be provided. Combinations of species can be placed in isolated waters such as ponds or reservoirs so that a natural food chain or varied kinds of angling can be furnished.

Aquaculture for the production of ornamental fish may not be limited by cost factors because the special challenge of many of the tiny colorful species is to raise them in captivity. Their value is determined in part by their scarcity and the difficulty of raising them. The search for colorful species with unusual social habits has stimulated extensive ichthyological exploration of little-known tropical waters and attempts to culture any colorful species that will live in small aquaria.

7.3 Selective Breeding

Farmers discovered long ago that plants and animals tended to resemble their parents and hence they chose superior individuals as brood stock. Archaeological studies frequently yield evidence of gradual improvement in domesticated species that was almost certainly due to selection.

Carp, which have been cultivated for 3000 or 4000 years in the Far East and at least 600 years in Europe, have been selected for such features as small head, high body, thick back, resistance to disease, suitability for various waters, and number of scales. Four fairly distinct morphological varieties exist: the original fully scaled carp, the mirror carp with a few large bright scales, the line carp with rows of scales only along the lateral line and dorsal fin, and the leather carp which is almost scaleless.

The development of trout breeding and culture during the middle of the nineteenth century was accompanied by much selection of brood stock, accidental and purposeful. Selection was accidental because the brood stock had to survive the rigors of confinement. It tended to become resistant to epidemic diseases, tolerant of artificial feeds and feeding practices, but poor at finding food and avoiding predators in the wild.

Controlled and scientific breeding programs in trout cultural stations have sought to develop strains that will resist disease, convert food to flesh efficiently, mature early, reproduce at various seasons, produce large numbers of eggs, and look attractive. Recently, at public trout hatcheries that rear trout for release in natural waters, some attempts have been made to select for good survival in the wild. Such an ability is, however, difficult to accomplish in a hatchery. Ideally, the ones that do survive in the wild should be used for breeders, but obtaining their progeny is rarely possible.

The process of selective breeding involves choosing superior parents for either crossbreeding or inbreeding. The relative merits of each have been subject to much argument. Crossbreeding tends to produce young that on the average have a higher growth rate than the young producd through inbreeding. It has been discovered, however, that deleterious characteristics occur occasionally through mutations that tend to be recessive. These characteristics can be eliminated from the population through inbreeding followed by rigorous selection to eliminate the individuals with undesirable characteristics. After several generations, "pure-bred" lines can be produced, which may then be crossed so that "hybrid vigor" or special combinations of desirable characteristics can be obtained.

Few fish breeding programs have developed closely inbred lines. The closest approach to a pure-bred trout is probably a stock of rainbow trout at the University of Washington in Seattle that has been inbred and selected by L. R. Donaldson for more than 30 years. He has trout now that tolerate poor water conditions at the hatchery, grow very rapidly on a rich diet, mature early, and reproduce effectively.

Fully controlled selective breeding of bivalve mollusks has not been possible until the recent development of methods of inducing spawning and caring for the larvae (see page 135). There has been, however, widespread transfer of stocks with desirable characteristics, such as the annual shipment of Japanese oyster spat to western North America, where the spawning of stocks is limited and uncertain. Also to be noted is the identification and transfer of disease resistant strains of the Atlantic oysters in eastern North America after epidemics decimated certain populations.

7.3.1 HYBRIDIZATION

A cross of two closely related species is frequently possible and may produce offspring of desirable characteristics. Such crosses occur fairly frequently in fresh water, especially among members of the Cyprinidae, the minnow family. The progeny are usually sterile and occasionally all of one sex. Sterile or monosexual fish are especially valuable for fish culture in cases of excessive breeding. *Tilapia* breeds at a length of only about 8 cm so effectively that it tends to overpopulate the ponds. Consequently, the discovery that a cross of an African male *Tilapia* with a Malayan female would produce only male offspring has immediate promise.

Crosses among the trouts, chars, and salmons have been tried and numerous hybrids produced, a few of which have desirable characteristics. The most successful cross to date in North America has been that of two chars, male brook trout and female lake trout. Their progeny, called "splake," are fertile. They have been distributed widely in eastern Canada and the United States to provide recreational fishing.

7.3.2 PHYSIOLOGY OF SPAWNING

Selective breeding of aquatic animals and the distribution of either eggs or larvae are much easier when man can intervene in the natural spawning process in ways that allow positive selection of the parents, collection and easy isolation of the young, and procurement of a supply of eggs whenever desired. Such intervention may be the only way in which mating can be accomplished between animals that tend to breed at different times or between varieties that will not mate naturally.

One of the great advantages of trout and salmon for aquaculture arises from the ease with which their normal spawning can be circumvented. The maturing fish are held in a pond in which they can be caught easily for determination of their reproductive condition. When they are "ripe," a few days before they would spawn naturally, the eggs and sperm can be pressed (stripped) from the bodies of the females and males into a pan in which the eggs are fertilized (Fig. 7.3). The fertilized eggs can withstand handling and shipping for a day or two before incubation in running water.

Additional control of rainbow trout spawning has been possible through selective breeding for a time of spawning. The wild stocks usually spawn in

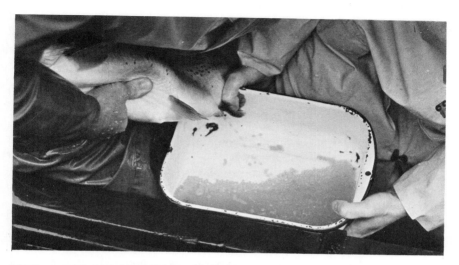

Fig. 7.3.   Stripping eggs from a large female rainbow trout (top) and stripping sperm from a male (bottom). (Photo by James O. Sneddon, University of Washington, Seattle, Washington.)

the spring. Small groups of trout have been induced to spawn earlier by holding them in warmer water, by lengthening their winter day artificially, and by feeding for more rapid growth. A few were found that could be stripped in the autumn and the process of selection continued. Now stocks are available that can be stripped in almost any month of the year so eggs can be supplied almost continuously for hatchery operations.

Usually carp must be allowed to mate naturally. They are placed in special spawning ponds, from which the larvae (fry) are collected after spawning and hatching. The larvae are transferred to nursery ponds or may even be shipped some distance. The control of the process has formerly been limited to placing selected breeders in proper spawning ponds in which the fry can be easily captured.

A greater degree of control of the reproduction has recently been obtained by injecting extracts of fish pituitary glands or other hormones into newly mature carp breeders. After such injections the breeders will spawn within a few hours under environmental conditions that might otherwise prevent spawning. Such practices have begun to make fry available during longer seasons and have permitted crosses of varieties that would not normally mate. Similar injections have been given to fish of numerous other species so that mating can be stimulated or eggs and sperm stripped.

Additional control of fish breeding would be possible if sperm could be stored for use when eggs became available. There have been numerous attempts to keep sperm by placing them in various physiologically neutral solutions and by quick freezing. Some success is apparent, and the establishment of sperm banks as a means of expediting breeding is likely.

The control of molluskan spawning has been especially difficult and important because they usually spawn during a brief period in summer and obtaining "sets" of larvae of the common commercial species is too uncertain. It has recently been discovered that spawning can be obtained any time of the year by conditioning the adults and then stimulating them to spawn. The conditioning is primarily a control of temperature—either raising artificially the temperature of the water in winter to about 20° C for 3 to 4 weeks for advancement of maturation, or cooling artificially the water in summer for delay of maturation. Spawning is stimulated by a sudden increase in temperature of a few degrees and by adding gonadal material from ripe individuals of the same species to the water. The development of the technique was essential for the further development of crossbreeding or inbreeding.

7.4 Nutrition

Almost any kind of confinement for an animal results in a lessening of the animal's ability to find food of the amount and kind required. It follows that keeping any kind of animal in confinement requires providing food for at least part of the animal's needs. In intensive culture of animals, the food that must be supplied is frequently the most critical and expensive item in the entire operation. Consequently, supplying the right food at the right time is a common determinant of the success of the culture.

Nutrition is the science of defining the interaction of an animal and its foods for the purpose of determining quantitatively the adequate food supply for different functions and stresses. It is a science that states the nutrient requirements in terms of components, such as lipids, amino acids, vitamins, and elements, compounds rations, develops feeding standards, and devises feeding methods. Nutrition involves the ways in which each nutrient is used by the body through the processes of digestion and metabolism, whether the nutrient is expended, stored, or secreted and whether it is wasted in the process. Nutrition involves the nutrients and body processes at all stages of life, especially the processes of growth and reproduction.

The feeding of aquatic animals in captivity remains largely an art, except for intensive feeding of fish such as trout and salmon. Some of the feeding of pond fish, such as carp, is done indirectly through fertilization of the pond and its management as a producer of live natural food for the fish. Captive mollusks are fed by placing them where they can filter their natural food from the passing water. Recent trials indicate that oysters will eat some starch from corn or wheat. The techniques of breeding mollusks require a supply of living food for the larvae, which usually must be grown in special cultures.

7.4.1 NUTRITION AND FEEDING OF TROUT AND SALMON

The early trout and salmon culturists discovered that the trout and salmon would grow on a diet of meat if their natural diet of zooplankton, insects, and fish was not available. Because meat was and is expensive, cheaper diets were sought both by trying cheaper kinds of meat and substituting grain products. Livers, lungs, hearts, and some kinds of fresh fish are excellent food for young trout or salmon, but these items are too expensive for most trout farmers. Consequently, a search for better and cheaper diets has been pressed principally at the U.S. Western Fish Nutrition Laboratory at Cook, Washington, and the Cortland Hatchery at Cortland, New York, operated under a cooperative agreement among the New York Conservation Department, the Fish and Wildlife Service of the U.S. Department of the Interior, and Cornell University.

Nutritionists proceed to determine essential nutrients by starting with two or more lots of standard healthy animals and feeding one lot a diet deficient in a single nutrient and simultaneously feeding another lot a standard diet. Faster growth in weight is the usual criterion of superiority, subject to conditions that diseases do not appear and the increase in weight is not largely deposited fat instead of muscle. Whole rations and various levels of either foods or nutrients are tried in similar feeding trials with similar performance criteria. The nutritionist may study further the utilization of nutrients by determining their wastage in digestion, places of storage in the body, rate of egestion or excretion. He will be especially alert for evidence of diseases associated either with deficiencies in diets or toxic effects. He will make studies of all of the above kinds at varying ages, especially when feeding starts, when rapid growth occurs, when reproduction takes place, and under varying climatic conditions.

Salmon and trout nutrition scientists[*] have identified many essential diet components, including numerous vitamins, amino acids, minerals, and fatty acids. They have concluded, for example, that trout can be reared on meat and meal mixtures at a rate of 350 g of protein and 2100 calories/pound[†] of trout produced. In this case about two-thirds of the calories are proteins.

Their findings have been developed into practical diets. One such diet for salmon and steelhead trout was developed largely by Oregon researchers

TABLE 7.1

OREGON PELLET FORMULA, JUNE 1963[a]

Meal Mix	Percent
Cottonseed oil meal (pre-pressed solvent extracted, 44% protein)	23.00
Herring meal (*Clupea harengus pallasi*)	21.00
Crab or shrimp solubles[b]	6.00
Wheat germ meal	3.60
Distiller's dried corn solubles	2.40
Vitamin premix (Table 7.2)	1.50

Wet Mix	Percent
Albacore tuna viscera (*Thunnus alalunga*)	20.00
One, or mixture of any of the following three:	20.00
Arrowtooth flounder (*Atheresthes stomias*)	
Pasteurized salmon viscera[c]	
Spiny dogfish (*Squalus acanthias*)	
Corn oil	1.80
Choline chloride (liquid, 70% product)	0.65
Antioxidant (Tenox IV)	0.05
	100.00

[a]Source: Hublou (1963).

[b]Composed of equal parts crab shell or shrimp meal and condensed fish solubles, dried together.

[c]Pasteurized 30 minutes above 140°F including 5 minutes above 180°F.

[*]Principally Arthur M. Phillips, Jr., at Cortland, New York and John E. Halver at Cook, Washington, and their associates.

[†] (Wet weight)

TABLE 7.2
OREGON PELLET VITAMIN PREMIX, JANUARY 1963[a]

Vitamin	Potency/pound (mg USP vitamin)
Ascorbic acid	23,586.7
Biotin	15.7
B_{12}	1.6
d-α-Tocopheryl acetate	9827.8
d-Calcium pantothenate	2830.4
Folic acid	188.7
Inositol	49,138.9
Menadione	157.3
Niacin	5031.8
p-Aminobenzoic acid	7862.2
Pyridosine hydrochloride	471.7
Riboflavin	1415.2
Thiamine hydrochloride	629.0
Total USP vitamin	101,157.0
Allowable carrier	352,433.0
Total (1 pound)	453,590.0

[a]Source: Hublou (1963).

(Hublou, 1963) and is used extensively in northwestern United States. Its formulation and proximate analysis are given in Tables 7.1–7.3. This diet is judged superior on the basis of growth, incidence of disease, and survival at sea. The pelleting system can be controlled so that food particles are provided with a minimum of wastage for fingerlings as small as 700 per pound.

TABLE 7.3
OREGON PELLET PROXIMATE ANALYSIS AND CALORIC
VALUE, JANUARY 1963[a]

Typical proximate analysis	Percent
Protein	34.8
Moisture	34.4
Carbohydrate (by difference)	16.5
Fat	7.5
Ash	6.8

Caloric value	Per pound
Digestible calories	1008

[a]Source: Hublou (1963).

7.4.2 LARVAL FEEDING

A critical time in the life of most aquatic animals is when they start to feed. They must have food of the correct size and shape. This food must supply the essential nutrients and be nontoxic. It must be abundant enough to be captured easily without an excessive expenditure of energy but not so abundant that it chokes the feeding animals or produces excessive waste products.

Many young fish that live pelagically in either fresh or marine waters start feeding when between 3 and 8 mm in length and when they have nearly consumed the food stored in the yolks of their eggs. For example, food eaten by larval sardines, which range in length from 4 to 6 mm, is usually copepod nauplii between 25 and 125 μ in diameter. They are estimated to require an average of 3.5 nauplii/hour at a water temperature of 14°C and twice that number at 19°C.

If such larvae are to be reared in captivity, they must usually be fed living food. Carp larvae of about 5 mm length eat plankton in their nursery pond which is managed to produce the plankton. Other captive larvae, such as plaice, shrimp, and many ornamental fish, are commonly fed the nauplii of the brine shrimp *Artemia*. Those are easily grown from *Artemia* eggs; the eggs are commercially available in most parts of the world.

Substitutes for living food for larvae have been extensively sought but are rarely satisfactory. Apparently the best have been the ground bodies of either fresh mussels or squid, but the requirements of a nearly continuous supply of nutritious particles of the correct size that will not contaminate the environment are difficult to meet. An added complication is the rapid growth of the larvae; they double in length in a few days and require different and larger foods.

Trout and salmon larvae are much larger than the larvae of most other bony fishes when they start to feed—20–30 mm in length. They are almost unique in accepting and growing readily on nonliving food, such as ground meat or ground liver. They do require, however, food particles of the correct size, frequent feeding, and amounts of food equal to those that they can consume so that there will be no excess to pollute their water.

Many of the larvae of mollusks start feeding at sizes of only 50–100 μ and have similarly critical food requirements. For example, larvae of the American oyster probably require naked flagellates for their first food but later are able to utilize phytoplankton with cell walls, such as *Chlorella*. The rearing of mollusks is now dependent on the culture of satisfactory living food supplies for all sizes. Such food has been supplied by growing combinations of algal species in fertilized seawater that has been treated with antibiotics for reduction of bacterial growth and with pesticides for reduction of infestations with zooplankton. Both of these chemicals must be used in quantities that will not damage the extremely sensitive molluskan larvae.

7.5 Diseases and Their Control

It is essential to expect the likelihood of disease among animals or plants in confinement to be worse than in their natural habitat. Closer spacing, greater bodily contact, and increased contact with body wastes make easier the physical transfer of disease organisms. Confined animals may also be more susceptible to disease because of increased stress due to either abnormal activity, abnormal diet, accumulation of wastes, interference with reproductive cycles, or increased fighting. Therefore, after intervening in the organisms' natural freedom, nutrition, or reproduction, it is usually necessary also to intervene on behalf of the organisms to protect them from disease.

The diseases of aquatic plants and animals are diverse and little known. They have been studied mainly from two viewpoints, first, by naturalists who have described the disease organisms found on wild hosts, and second, by pathologists who have described and tried to control the diseases of captive plants and animals. A large proportion of the latter have been concerned with the diseases of trouts and carps which are, therefore, probably better known than the diseases of any other groups of fish. Fortunately, the common diseases of trouts and carps are also common among many other kinds of fish.

7.5.1 IDENTIFICATION OF DISEASE ORGANISMS

Animals and plants commonly harbor many associated organisms which may be harmless at times, occasionally pathogenic, or regularly pathogenic. It is not easy to determine the causal agent of a specific disease and usually the identification of a causal agent depends on satisfying the postulates of Robert Koch, an early microbiologist. These are in essence:

1. The causal agent must be associated in every case of the disease as it occurs naturally.
2. The agent must be isolated in pure culture.
3. When the host is inoculated with the isolated organisms under favorable conditions with suitable controls, the characteristic symptoms of the disease must develop.
4. The causal agent must be reisolated and identified as that which was first isolated.

The pathogenicity of an organism is always subject to a combination of factors. The organism must of course be present; the host must be susceptible; the organism must enter the host and establish itself after overcoming any defenses; the organism must emerge again to be transferred to a new host; and the organism must survive both in the host and in the environment between hosts. When an organism kills its host without emerging, it too dies.

Thus the more successful disease organisms are those that do not seriously damage their hosts and that, by some means, tolerate easily the transfer between hosts.

7.5.2 KINDS OF DISEASES

Diseases are commonly classified according to the primary causal agent or factors, but it must be kept in mind that an organism weakened by one disease is often subject to others. Consequently, identification of the primary cause may be uncertain or treatment may be necessary for several diseases. A few of each of the several kinds of diseases are mentioned below.

Environmental. Aquatic animals in confinement depend to a large degree for their good health on suitable water. They tolerate considerable variations in salinity, pH, temperature, dissolved gases, etc., but when exposed to less then optimum factors, they become susceptible to disease. Excessively acid water (pH < 5.0) and excessively low temperature may seriously weaken some kinds of carp. One form of the disease, commonly called "popeye" in trout, is caused directly by water supersaturated with nitrogen. The gas comes out of solution as a bubble in the eye socket and causes the eye to protrude. Captive animals are also commonly subject to physical injury from nets or sorting equipment, as well as from predators or fighting among themselves. Salmon are occasionally subject to sunburn.

Dietary. Diseases that are caused directly by deficiencies in the diet include: anemia, due to a varied vitamin deficiency; goiter, due to a deficiency of iodine; blue slime disease of trout, due to a deficiency of biotin; and dietary gill disease, due to a deficiency of pantothenic acid. A disease due to overfeeding with poorly balanced diet is lipoid degeneration of the liver, to which rainbow trout are especially susceptible.

Virus. Few virus diseases have been positively identified, but many infectious illnesses are thought to be caused by viruses. Their identification is especially difficult because of the problems of isolating and identifying them and the prevalence of secondary infections by bacteria which influence the course of the disease. One disease of trout that is caused by a virus is infectious pancreatic necrosis.

Fungus. Fungi or water molds of the Saprolegniaceae are common in all fresh water and frequently infect fish of many species. They appear as cotton-like growths on the skin, but in many instances they are secondary invaders of a lesion caused by injury or bacteria. They also appear quickly on dead bodies or eggs. Other fungi may invade the internal organs of fish, such as the air bladder and intestine. Still other fungi, especially *Dermocystidium*, are probably the causative agents of the worst-known infectious diseases of oysters. *Dermocystidium* attacks the body of the oyster; other fungi may attack the shell and the attachment of the adductor muscle.

Bacteria. Numerous bacterial diseases of fish have been identified. Many are evidenced by external lesions on either gills, body, fins, or caudal peduncle. Furunculosis, which appears as boillike or ulcerlike skin lesions, is caused by the bacterium *Aeromonas.* An ulcer disease of trout is caused by *Hemophilus*, a small rod-shaped bacterium (Fig. 7.4). Bacterial gill disease, evidenced by excess irritation and necrosis of the gills, is caused by several species of *Myxobacteria.* Other *Myxobacteria* cause columnaris, a disease

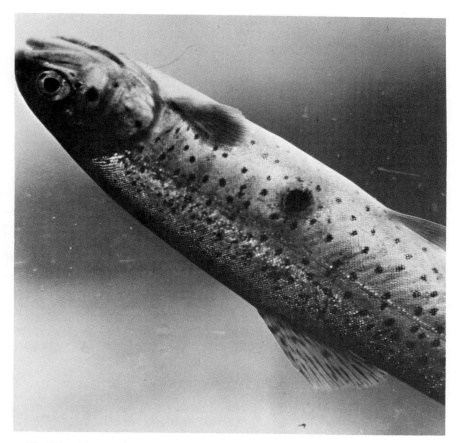

Fig. 7.4. A lesion of ulcer disease on a small rainbow trout. (Photo by Frederick F. Fish, courtesy of U.S. Bureau of Sport Fisheries and Wildlife.)

marked by yellow open sores on the head and body. Other infectious internal diseases caused by bacteria include kidney disease and infectious dropsy which is one of the serious diseases of carp.

Internal Parasites.   One should expect to find internal parasites in any animal that is examined, whether it appears healthy or not. They occur in bewildering variety; most of them are relatively harmless, but a few cause serious trouble for their host. Especially prevalent and rarely harmful are the intestinal worms, such as tapeworms (Cestoda) and the thorny headed worms (Acanthocephala). More serious when they occur in the musculature are the roundworms (Nematoda).

Most of the internal parasitic worms have a life cycle involving occurrence in a vertebrate host, transfer to an invertebrate host, sometimes transfer to another invertebrate host, and transfer back to the vertebrate host (Fig. 7.5). The invertebrate hosts are commonly planktonic Crustacea. Larval fish commonly feed on planktonic Crustacea, so infection of young fish occurs very early in life. A secondary invertebrate host is frequently a snail. Occasionally a secondary vertebrate host is a fisheating bird or another fish.

The parasites enter the host along with the food, but some have free-swimming stages, which penetrate the skin or gills. Many can also migrate within

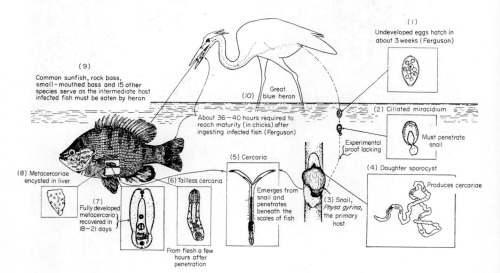

Fig. 7.5. Diagram of the life cycle of a white liver grub, *Neodiplostomum multicellulata*, that infects many species of the family Centrarchidae in North America (1, 4, 5, after Miller) (7, after Hughes) [After G. W. Hunter III (1937). "Parasitism of Fishes in the Lower Hudson Area," A Biological Survey of the Lower Hudson Watershed, pp. 264–273. New York State Conserv. Dep. Albany, New York.]

the host at some stage by penetrating the walls of the intestine or blood vessels.

Other internal parasites of fish include protozoans, such as *Myxosporidia* and *Octomitis*. The latter is a small flagellate that causes infectious disease in trout and salmon. A haplosporidian, *Minchinia*, has caused drastic mortalities in oyster populations. Also serious at times are *Trypanosomes*, which live in the blood of fish and are frequently transferred by leeches (Hirudinea).

External Parasites.   External parasites, like the internal ones, also appear in bewildering variety and usually cause little harm to the host. The parasites of fish that may cause serious disease include larval mollusks, parasitic crustaceans, parasitic worms, and protozoans. The larval mollusks are the drifting larvae of certain clams. They attach temporarily to the gills and body of fish and may be harmful if very abundant. The crustaceans, frequently called

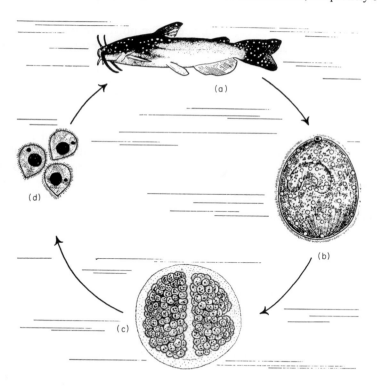

Fig. 7.6.   Life cycle of *Ichthyophthirius multifilus*. (a) Parasites in skin of catfish; (b) free-swimming parasite after leaving fish; (c) encysted parasite which has divided into a large number of young; (d) young parasites which have left the cyst and are searching for a fish host. (After Davis, 1965.)

sea lice, can attach themselves to the skin, either with a sucking disc or with the head, which actually penetrates the flesh. These, too, are harmful only when unusually abundant. More likely to be abundant and troublesome are the external trematode worms of the genus *Gyrodactylus* and closely related genera. These worms, about 1 mm long, attach to gills and fins by means of hooks. They are likely to be especially dangerous to young carp and young trout. Unlike the internal parasites, this animal reproduces hermaphroditically without an intermediate host. The young even have developing eggs or young before birth, and one adult may carry three successive generations within it.

Other troublesome external parasites of fish are the protozoans *Costia* and *Ichthyophthirius* (Fig. 7.6). The latter is widespread among both captive and wild fish and causes serious mortality when abundant.

Very different are the external parasites of oysters. Many plants and animals live on oyster shells with no effect on the oyster, but some invade the shell or body with serious consequences. The boring sponges, clams, and mud worms (Annelida) invade the shells and can seriously weaken the oysters.

7.5.3 DISEASE CONTROL

The primary control of diseases of fish or mollusks is, like preventive medicine, based on good nutrition, good water, sanitation, and avoidance of overcrowding. The food must supply essential nutrients but should not transmit disease organisms. Organisms can be killed in trout foods by pasteurization. In carp ponds, where the young eat live foods, organisms can be killed by drying and sterilizing the pond before stocking. Food must also be available to all individuals; otherwise the large fish may eat well while the small fish starve.

All fish diseases are water borne, and any fish in the water supply is likely to carry disease. Few water supplies are free of fish, so a source of infection must be expected. When epidemics occur, they are much easier to control when the ponds are arranged in parallel instead of in series. In addition, the operator can avoid the spread of disease by sterilizing everything that comes in contact with the water in an infected pond.

Some additional chemical treatments will help when disease does strike. Chemicals are given in dosages or concentrations for periods that harm the disease organisms more than they do the host. Some bacterial diseases, such as furunculosis, are controlled by feeding antibiotics, such as sulfamethazine or Terramycin, in the food. On the other hand, diseases caused by external parasites are frequently controlled by the addition of chemicals to the water. These are poisons such as organic mercury compounds, formalin, copper sulfate, malachite green, and quarternary ammonium germicides. The chemical is administered in baths for limited periods at concentrations carefully controlled so that harm to the fish is minimized.

When dietary, sanitary, and chemical methods fail, as they do frequently in our present state of knowledge of fish disease treatment, the principal alternative is to find disease-resistant strains. Carp and rainbow trout have been bred for resistance to some of the troublesome diseases. Oyster populations that were almost eradicated by epidemic have been restored by breeding the survivors, which presumably were resistant to disease.

It will be obvious from the foregoing discussion that the knowledge of disease organisms and pathology of aquatic animals is woefully small. Much more scientific study will be needed to bring our knowledge close to that of the diseases of domestic land animals.

REFERENCES

Major general works on aquaculture of fish or mollusks include the following:

Bennett, G. W. (1962). "Management of Artificial Lakes and Ponds." Van Nostrand-Reinhold, Princeton, New Jersey.
Davis, H. S. (1953). "Culture and Diseases of Game Fishes." Univ. of California Press, Berkeley and Los Angeles.
Food and Agriculture Organization. (1967). *FAO World Symp. Warm-water Pond Fish Cult., 1966* A collection of 123 papers.
Hickling, C. F. (1962). "Fish Culture." Faber & Faber, London.
Hora, S. L., and Pillay, T. V. R. (1962). Handbook on fish culture in the Indo-Pacific region. *FAO Fish., Biol. Tech. Pap.* **14**, 1–204.
Leitritz, E. (1959). Trout and salmon culture. *Calif. Dep. Fish Game, Fish Bull.* **107**, 1–169.
Loosanoff, V. L., and Davis, H. C. (1963). Rearing of bivalve molluscs. *Advan. Mar. Biol.* **1**, 1–136.
McNeil, W. J., ed. (1970). "Marine Aquiculture." Oregon State Univ. Press, Corvallis.
Medcof, J. C. (1961). Oyster farming in the maritimes. *Bull., Fish. Res. Bd. Can.* **131**, 1–158.
Ryther, J. H., and Bardach, J. E. (1968). "The Status and Potential of Aquaculture," Vols. 1 and 2. Amer. Inst. Biol. Sci., Washington, D.C.
Schaperclaus, W. (1961). "Lehrbuch der Teichwirtschaft." Parey, Berlin (French edition by Vigot Frères, Paris, 1962).
Schuster, W. H. (1952). Fish culture in brackish water ponds of Java. *Indo-Pac. Fish. Counc., Spec. Publ.* **1**, 1–143.

Important contributions to special aspects of aquaculture or the basic sciences include the following:

Cortland Hatchery Reports. (1932–1971). Bur. Fish Cult., New York State. (Title varies.)
Dogiel, V. A., Petrusheushi, G. K., and Polyansk, Yu I., eds. (1958). "Parasitology of Fishes." Leningrad Univ. Press (English transl. by Z. Kabata, Oliver & Boyd, Edinburgh, 1961).
Fisher, R. A. (1965). "The Theory of Inbreeding," 2nd ed. Academic Press, New York.
Gordon, M. (1957). Physiological genetics of fishes. *In* "The Physiology of Fishes" (M. E. Brown, ed.), pp. 431–501. Academic Press, New York.

Hublou, W. (1963). Oregon pellets. *Progr. Fish Cult.* **25**, 175–180.

Mitchell, H. H. (1963, 1964). "Comparative Nutrition of Man and Domestic Animals," Vols. 1 and 2. Academic Press, New York.

Schaperclaus, W. (1954). "Fischkrankheiken," 3rd ed. Akademie-Verlag, Berlin.

Shanks, W. E., Gahimer, G. D., and Halver, J. E. (1962). The indispensable amino acids for rainbow trout. *Progr. Fish Cult.* **24**, 68–73.

Sindermann, C. J., and Rosenfield, A. (1967). Principal diseases of commercially important bivalve mollusca and crustacea. *U.S., Fish Wildl. Servo., Fish. Bull.* **66**, 335–385.

Snieszko, S. F., ed. (1970). A symposium on diseases of fishes and shellfishes. *Amer. Fish. Soc., Spec. Publ.* **5**, 1–526.

Wood, J. W. (1968). "Diseases of Pacific Salmon, their Prevention and Treatment." Hatch. Div., Washington State Dep. Fish.

$$8$$

Fisheries and Their Methods

8.1 History of Food Fisheries

Fishing started long before man existed on earth. A number of the larger modern mammals, such as bears and raccoons, catch fish and crustaceans or dig for mollusks, as presumably their ancestors did. Early man located his habitation near good hunting or good fishing and used his arrows or spears or traps to capture fish as well as birds or mammals. When man began to domesticate plants and animals commonly, he continued to hunt and fish to augment the production of his fields, especially during seasons when no crops were available.

Fishing was closely linked to water transportation. The rafts and canoes that were used for fishing could also be used for transport. The port that became a center of commerce also became a center of fishing, and frequently the fish were an important part of the commerce.

Most of the kinds of fishing gear now in use were used long before the birth of Christ. A line was attached to an arrow to retrieve the fish. A barb was added to the spear to form a harpoon. Hooks were fashioned from either wood, bone, or metal. Traps were made from rocks, brush, or reeds. Nets were made from fibers and used to dip fish, to surround them, or to entangle them. Many of the devices indicate a remarkable adaptation of the materials available to the behavior of the fish being sought.

Even the methods of aquaculture are ancient. Agriculture, which started 8000–10,000 years ago, included some aquaculture as farmers who had ponds began to keep fish alive in them. They learned to transport fish and eggs in order to stock the ponds and learned further that animal and plant wastes could be used to fertilize the pond. Farm and village ponds for fresh-water fish were common in the eastern Mediterranean area from 2000 to 1000 BC. Brackish water ponds were used by the early Romans to hold oysters.

The old practices of both fishing and aquaculture persist with little change in many parts of the world today. Villages exist along coasts and waterways of Asia, Africa, and America in which the fishermen still either paddle or pole their boats (Fig. 8.1). They use gear fashioned of local materials (Fig. 8.2). They either consume the fish fresh or dry it over crude fires (Fig. 8.3). Frequently the villages have been located where they are because of the

Fig. 8.1. A fisherman on Madagascar returning to his village of Anororo. (Photo by P. Pittet, courtesy of FAO.)

Fig. 8.2. A Yemeni fisherman making a fish trap. (Photo by H. Kristjonsson, courtesy of FAO.)

fish supply, and they have grown to a size sustainable by the fish within reach of the fishing boats. In consequence, they are frequently in remote areas with little contact with other people. The fishermen are mostly illiterate; the fishing practices are a part of the traditional patterns of family and village life that continue with strong resistance to change. In these villages fishing is carried on together with primitive agriculture. The average fisherman in such a village produces only 1 ton or less of fish per year.

The development of modern fishing practices began with improvements in sailing craft, navigation, and charts. Sailing vessels enabled men to carry larger nets, to fish much farther from home and, more important, to bring home the catch if they could preserve it. They could sail thousands of kilometers to catch whales for oil or cod for salting. The development of long range vessels and large gear was a marked departure from the primitive practices of the fishing village. Fishing was conducted mostly out of a seaport—an urban rather than a rural base. Furthermore, the men who were engaged either became full-time fishermen or, if they had auxiliary employment, tended to work at urban trades.

Fig. 8.3. Kilns for smoke drying fish on the Ivory Coast. (Photo by A. Defever, courtesy of FAO.)

With the development of power boats during the latter part of the nine-teenth century came the use of much larger gear. Giant trawl nets could be pulled along the bottom at great depths. Immense seines could be dropped quickly around schools of fish. The power provided other impor-tant advantages; the boats could operate more safely and return to port on a schedule with fresh fish, preserved by ice.

Other improvements followed rapidly. Trawl nets were modified with great rollers to carry them over relatively rough bottoms. Electronic aids to navigation facilitated the location of fishing grounds. Later, they were modified to find fish. Artificial fibers were developed that were stronger, lighter, and more resistant to rotting. Facilities for mechanized refrigera-tion of the catch were added to vessels. These were designed at first to chill the catch, later to freeze it and carry it for months. More and more equip-ment for the handling and processing of fish was added to vessels, and eventually they became floating factories able to fish in any ocean and to process, package, preserve, and transport the catch (Fig. 8.4).

Fig. 8.4.   A Soviet stern trawler of the Mayakovsky class. It displaces about 3700 tons, is 85 m long overall, carries a crew of about 110, pulls a trawl 38 m wide, and freezes up to 30 tons of fish daily. Several hundred trawlers of about this size are operating on the major trawling grounds of the world. (Photo by U.S. National Marine Fisheries Service.)

8.2   Recent Trends in Production

Statistics on total fishery catches for any large parts of the world are nonexistent for periods prior to about 1900 because much of the fishing was done from primitive villages. During the 18th and 19th centuries the better-developed countries in Asia, Europe, and North America instituted systems for collecting statistics on the production of major commercial products, such as whale oil and salt fish. Worldwide catch statistics were not collected until the formation of the Fishery Section of the Food and Agriculture Organization of the United Nations. The Organization began to develop a system for collecting fishery statistics worldwide immediately after World War II. They started with the year 1948 and tried to procure comparable data for 1 prewar year, 1938. The task turned out to be so formidable that they did not attempt a complete collection for 1949, but beginning with 1950, they have prepared estimates of total world fish production and trade by species, country, and major fishing areas. These are published in the "Year-books of Fishery Statistics," of which there are 28 volumes at this time.

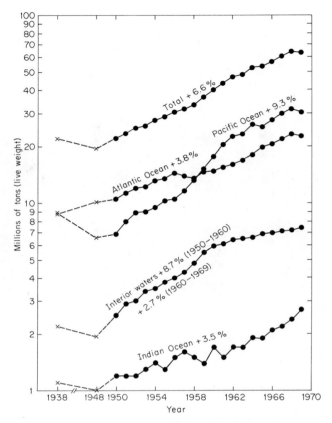

Fig. 8.5.   Recent trends in total world fish catch by waters. (Source: FAO.)

The trends shown in these statistics (Figs. 8.5, 8.6, 8.7) need to be judged in relation to the trends in overall food supplies, which increased annually from 1950 to 1966 at an average rate of a little over 3%. This rate for the world at large was only slightly greater than the rate of population increase, so the food supply per person increased only slightly. Of course fish is a high-protein food and is especially important for improving diets where protein supplies are deficient, so the trends need to be evaluated also with consideration of the fact that fish production is about one-fourth of the world meat production from terrestrial mammals and birds.

Most noteworthy is the steady increase in total world catch (excluding whales) averaging 6.6%, or more than double the rate of increase in terrestrial food products. Not only is the upward trend in the total remarkably steady from year to year, but the different continents and waters all

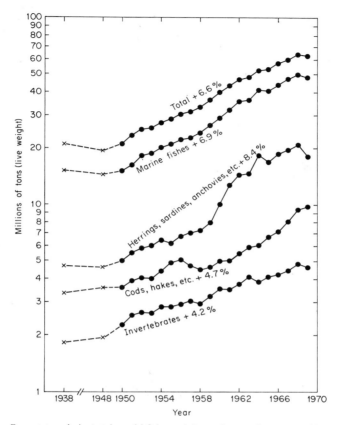

Fig. 8.6.   Recent trends in total world fish catch by major species groups. (Source: FAO.)

show upward trends in excess of 3% except North America and, recently, the inland waters (the last is questionable because of the large component of uncertain data from interior Asia).

The greater rate increase appears in the Pacific Ocean off South America, among the herring, sardine, and anchovy group of species because of the exceptional development of the anchovy fishery off Peru and Northern Chile. These fish and many other small fish are being sought increasingly for drying into fish protein concentrates. In contrast, the production of the Northern Hemisphere, which in total quantity has formed the greatest part of the total world fish catch, has been increasing much less rapidly.

Not only South America, but Africa, Asia, and the USSR show rates of increase greater than Europe and North America. They reflect the much greater rates of increase in fish production in the countries with less-

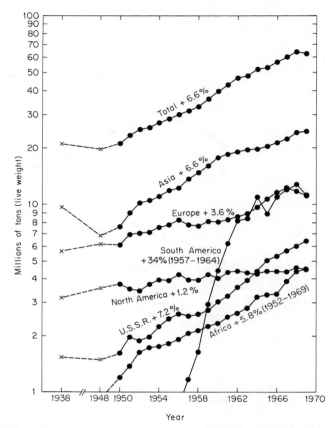

Fig. 8.7.   Recent trends in total world fish catch by continents. (Source: FAO.)

developed or centrally controlled economies rather than in the countries with well-developed economies. Most of the latter, including the countries of western Europe, Canada, Japan, Australia, New Zealand, and the United States, are increasing their fish production slowly if at all. In the United States the total fish production has remained at nearly the same level for more than 30 years.

8.3 Modern Fishing Systems

Capturing fish and transporting them from the sea to a fishing port requires an integrated fishing unit; a vessel or fleet that can carry men to fishing grounds where they will operate gear to catch the fish and partially process and preserve them immediately and then bring them to port in good condition. The vessel(s) must be safe, seaworthy, and habitable. The vessel(s) must be equipped to carry, set, and retrieve the fishing gear. The gear must be suitable for the species of fish sought and the grounds being fished. The vessel(s) must be supplied to travel to and from the fishing ground and to operate while fishing. It must be arranged to permit the necessary stowage and transport of the catch. The vessel(s) must also be provided with space to handle the fishing gear and the catch, with winches necessary to operate the gear and with means of stowing the gear between fishing operations.

The primary determinants of vessel size and design are the distance to the fishing grounds and the kind of gear to be used. On nearby grounds in sheltered waters, small vessels can make short trips to land with fish fresh

TABLE 8.1
CLASSIFICATION OF PRINCIPAL FISHING GEARS

1. Hand diggers and collectors
     Manually operated rakes or digging equipment for mollusks,
       crustaceans or other burrowing animals
     Trained mammals or birds
2. Dredges
     Power-operated rakes or excavators for mollusks
3. Spears
     Hand-held spears on a shaft
     Harpoon (shaft with detachable head on a line) for large fish
     Explosive harpoon for whales
4. Stupefying aids
     Poisons, explosives, or electricity
5. Hooks on lines
     Hand-held casting lines
     Trolling lines towed from boat
     Set lines (long lines) with many hooks on surface or bottom
6. Stationary entangling nets
     Gill nets with single wall of mesh
     Trammel nets with multiwalled mesh
7. Stationary enclosures
     Large net enclosures, corrals, true traps for coastal fish
     Small net enclosures, fykes, bag nets for river fish
     Brush or rock enclosures for coastal or river fish
     Rigid pots for eels, crustaceans
8. Mobile enclosing nets
     Trawls towed along bottom or at mid-depths
     Seines pulled on bottom toward a fixed point
     Purse seines, floating with purse line to close bottom of seine
     Falling nets, cast nets
     Lift nets

or even alive. In coastal and regional areas, larger vessels can travel up to about 1500 km and return with catches preserved in ice for sale on the fresh-fish market. When trips need to be longer than about 15 days, however, preservation in ice is usually not satisfactory, and the vessels must be equipped for either freezing, canning, salting, or drying of the catches. Thus, fishing distant grounds requires that vessels be designed for much longer trips and for full preservation of the catch. These requirements may be met by large, single vessels between 70 and 80 m in length or by fleets in which small catcher vessels will be serviced and resupplied from motherships in which the catches will be processed.

Fishing gear has evolved over the centuries as observant and ingenious men have adapted the materials at hand to the capture of fish. A bewildering variety of gear exist that can be classified in many ways (Table 8.1). Any of these gear can be highly efficient under appropriate circumstances, but the prevailing tendency in modern operations is to use active gear in which power is applied in order to take the gear to the fish instead of stationary gear which the fish must find. The great majority of fish are now captured by otter trawls (Fig. 8.8) and by purse seines (Fig. 8.9), but important quantities are taken by gill nets (Fig. 8.10), traps (Fig. 8.11), and hooks (Fig. 8.12).

The earliest trawls were cone-shaped nets that were towed along the

Fig. 8.8. Diagram of an otter trawl.

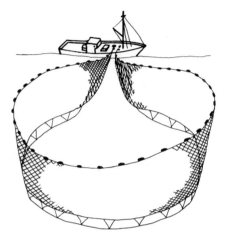

Fig. 8.9.   Diagram of a purse seine.

bottom by sailing vessels. The front of the net was held open by a rigid frame, a beam at the top, and metal frames with runners at each end. This "beam trawl" was limited in size by the length and weight of a beam that could be handled on the vessel. It was soon found to be possible to hold the net open by kitelike "otter boards" that were weighted to slide along the bottom. With otter boards and with powered vessels it was possible to handle much larger nets, or "otter trawls." These larger nets were modified with rollers on the bottom of the net that would pass over rough bottom, with floats and fully cut tops to increase the vertical dimension. One modification that produced a great increase in efficiency was the Vigneron-Dahl gear—a

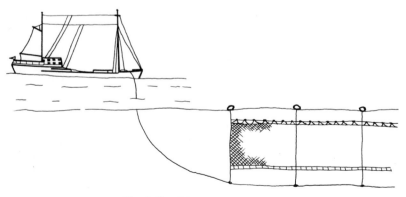

Fig. 8.10.   Diagram of a gill net.

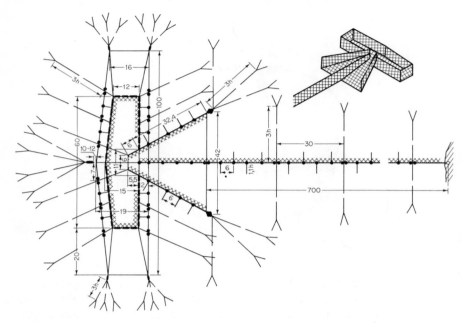

Fig. 8.11.   Diagram of a floating salmon trap used in the sea off eastern Siberia. Dimensions of the netting are in meters, and of the anchor lines in multiples of the depth (h) of water. [Source: V. S. Dolbish, ed. (1958). Oyidiya rybolovstva dalnevostochnogo basseina. Pishchepromizdat, Moscow.]

linkage between the otter boards and the net that permitted the boards to be operated far from the net and the herding effect of the towing warps to be utilized while still allowing the use of power to bring the net on board. More recent modifications have been square openings maintained by a system of weights and otter boards, resulting in a midwater trawl that can be operated independently of the bottom.

Seines are flat nets used to encircle fish, fitted with floats on the top and weights on the bottom. Beach seines are set on the bottom by hand or from small vessels and pulled back up on the beach. Purse seines are fitted with an excess of floats and are set to hang from the floats at the surface regardless of the depth of the water. They are fitted also with a purse line in the bottom so that the bottom can be closed after they have been set in a circle around a school of fish. Modern purse seines are used for pelagic schooling fishes such as tuna, herring, mackerel, salmon, or anchovy. Some are very large, often as much as 1000 m long and 200 m deep, and require a large vessel rigged especially to handle them. The use of modern fibers has revolutionized the purse seine fishery by permitting larger nets that sink more rapidly

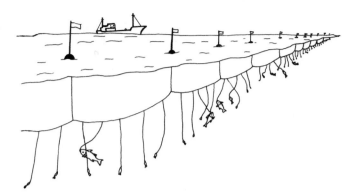

Fig. 8.12.   Diagram of a long line, one of the kinds of hook gear.

and can be stowed while wet in minimum space without deterioration.

Gill nets are flat wall like nets constructed of fine threads that entangle fish. They are fitted either with an excess of floats over leads to hang from the surface or with an excess of leads over floats to float upward from the bottom. They are set to remain in position where fish congregate. The use of modern fibers and filaments, especially monofilament nylon, in the construction of gill nets has increased their efficiency. They are especially favored by small-boat fishermen in sheltered waters because they can be operated by hand or with simple hauling machinery.

Traps are made in great variety, but all are stationary and all depend on guiding animals in through openings that are difficult or impossible to find in order to get out. Pots for shrimps, crabs, or lobsters are simple rigid traps to which the animals are enticed by bait. At the other extreme of size are ocean fish traps that may have a lead net more than 1000 m long. Other traps of intermediate size and many names, such as bag nets or fyke nets, have been devised to take advantage of the habits of many animals in estuaries, rivers, and lakes.

Hooks are used in many ways, but the most productive are the set lines or long lines used for large animals that do not school. These consist of a main line stretched between floats or anchors, to which are attached short "dropper" lines, at intervals, with baited hooks. Large, strong lines are used on the sea bottoms for sharks and halibut and near the surface in warm waters for large tunas, marlins, and sharks.

Despite the fact that all of the common methods of fishing are very old, there has been a real technical revolution in the development of vessels, the improvement and enlarging of fishing gear, the use of better methods of preservation, the use of new, strong, nonrotting fibers, and the use of

TABLE 8.2
GENERAL PRODUCTION LEVELS OF VARIOUS FISHING PRACTICES

Catch (tons/man/year)	Fishing method
1	Primitive fisheries in which lines, traps, spears or nets are fished from manually operated boats
3	Primitive fisheries as above with small power boats
10	Fishing from small coastal vessels with lines, trawls or gill nets for highly valuable fish
30	Fishing from medium to large vessels with lines, trawls or purse seines for moderately and highly valuable fish.
100	Fishing with the best modern trawlers and purse seiners for moderately and highly valuable fish
300	Fishing with modern purse seiners for fish of low value, such as pilchard or anchovies.
1000	Fishing with most efficient purse seiners for anchovies

electronic equipment for navigation and finding fish. The result is a huge increase in the productivity per man when all of the modern equipment and methods are utilized. The dramatic difference in levels appears (Table 8.2) when different kinds of present-day fishing are compared. These are shown as general levels because good overall statistics are scarce, but it is noteworthy that the average production for hundreds of thousands of primitive fishermen in the little-developed countries is 1 ton or less per man per year, whereas the average production per man in Iceland is nearly 100 tons.

8.4 Preservation and Processing

Fish are extremely diverse in kind. Some serve as the mainstay of the diet for the hungry, others as appetizers for the gourmet. The hundreds of kinds that are different differ markedly in texture, oiliness, boniness, and flavor in ways that require careful attention to methods of preservation and preparation. Such special individual care has been given in the home or restaurant by the cook but it is gradually being provided in the processing plant as an increasing proportion of our fish are prepared before retail sale.

The average gross chemical composition of fish flesh (exclusive of viscera and skeleton) is about 75% water, 19% protein, 5% oil, and 1% ash or inorganic compounds. Whole fish have a somewhat higher proportion of oil because of the oil in liver and viscera and of ash because of the minerals in the bones. The principal variation among species is in oil content, which is less than 1% in lean fish, such as cod and hake and more than 10% in oily fish, such as sturgeon, some salmon, mackerel, and herring. Oil content also

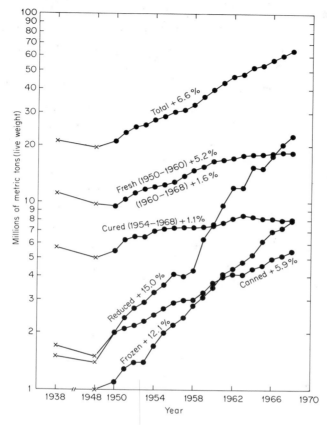

Fig. 8.15.   Disposition of world catch. (Source: Food and Agriculture Organization of the United Nations, "Yearbook of Fishery Statistics," Vol. 27. FAO, Rome, Italy.)

comprise about one-third of the total world production. The next most rapid increase has been in frozen fish, 12%, and canned fish, 5.9% annually. On the other hand, the production of fresh fish and cured fish, the traditional products, has increased more slowly, especially in the late 1960's, when they seemed to be leveling off.

8.5   Consumption and Distribution

People eat fish when it is available because they like it, they are accustomed to it, or they find a religious significance in it. They may also feed it to their animals or use nonfood products of various kinds. In practically all cases, they choose it from among others foods or products on the basis of their preferences and its relative price. Rarely do people eat fish because they think it is good for them.

Fish makes a relatively tiny contribution to the world calorie supply, only about 1%, and a somewhat larger contribution to the world protein supply, which comes from meat, milk products, eggs, grains, nuts, and pulses. But the principal value of fish is in the quality of its protein. Like meat, milk products, and eggs, it supplies a relatively high proportion of lysine, threonine, and tryptophan, amino acids that are essential in the human diet but scarce in the plant proteins.

Since fish has been increasing in supply rapidly, it has been considered especially important in alleviating the malnutrition caused by protein deficiency. This, along with calorie deficiency, is especially serious in the developing countries where protein is scarce and expensive. Those worst afflicted are children because their requirement for both calories and protein is from one-third to one-half that of an adult even at very early ages (Fig. 8.16) and they are frequently fed diets high in starches and sugars. The consequence is a death rate for ages 1–4 of 10–30/1000 in developing countries as compared to a death rate of about 1/1000 in the developed countries. Protein deficiency disease, kwashiorkor, is believed to affect up to half of the young children in some countries. This disease occurs even in countries whose total supply of protein is more than adequate because high-protein food is unevenly distributed, because of cost, ignorance, or deeply ingrained dietary habits.

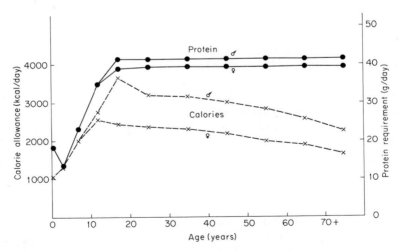

Fig. 8.16. Caloric allowance and protein requirement per person per day by age. (Source: *FAO Nutr. Ser.* **15**; *World Health Organ., Tech. Rep. Ser.* **301**. FAO standard body weights.)

TABLE 8.3
SUPPLIES PER PERSON IN GRAMS PER DAY OR FISH AND TOTAL ANIMAL PROTEIN IN SELECTED COUNTRIES[a]

	Fish (edible weight)[b]	Total animal protein (dry weight)
Western Europe		
France	21	59.5
Italy	16	33.2
Norway	55	50.5
Portugal	62	30.4
United Kingdom	26	52.7
North America		
Canada	19	63.0
United States	14	65.1
Latin America		
Argentina	8	47.9
Brazil	6	18.7
Ecuador		14.2
Near East		
Israel	18	39.7
Jordan	2	9.6
Turkey	6	15.9
United Arab Republic	14	12.6
Far East		
India	3	5.7
Japan	76	24.6
Pakistan	4	9.5
Philippines	45	15.9
Africa		
Ghana	22	11.1
Kenya	3	12.1
South Africa	24	31.5

[a]Source: Food and Agriculture Organization of the United Nations. (1967). "The State of Food and Agriculture." FAO, Rome. Italy. Most data are for 1965–66, but, when these were lacking, earlier data were used.
[b]The edible weight is about 20% protein.

The consumption of fish may be judged on its contribution to the supply of animal proteins (Table 8.3). In some countries with moderate supplies of total animal protein, such as Japan and the Philippines, fish may provide about two-thirds of the total. In the developed countries of Western Europe and North America with large supplies of animal protein, fish may supply only 2–10%. In the developing countries of Asia, Africa, and Latin America, fish may supply an important percentage of a very small supply of animal protein.

TABLE 8.4

INTERNATIONAL TRADE IN FISH AS A PERCENTAGE OF
TOTAL WORLD PRODUCTION[a]

Item	Percent in International trade
Total fish (1950)	21
Total fish (1960)	31
Total fish (1968)	45
Chilled or frozen fish (1968)	35
Cured fish (1968)	16
Crustaceans and mollusks (1968)	33
Fish products and preparations (mostly canned) (1968)	19
Fish oils and fats (1968)	64
Meals, solubles (1968)	71

[a]Source: Food and Agriculture Organization of the United Nations (1968). "Yearbook of Fishery Statistics," Vol. 27. FAO, Rome, Italy.

The consumption of fish in various countries is, as should be expected, extremely uneven. It is very important in the diet in a few countries, non-essential in others, and could be important in some countries in which animal protein is scarce.

Another indicator of the importance of fish is its changing role in international trade. Before World War II fish were consumed predominantly in the country in which they were landed. Even in 1950 only 21% of the production entered international trade (Table 8.4). The percentage has increased rapidly to 45% in 1968. The items that are now most important in international trade are the fish meals and solubles, fish oils and fats, crustaceans and mollusks, and chilled or frozen fish. The items fall into two major groups; the "industrial" products, which are produced in large quantity in only a few places, and higher-priced items, such as frozen fish fillets and shrimp. One or both groups are important sources of foreign exchange for many developing countries; for example, about 50 countries export shrimp to the United States.

The total international trade in fishery products in 1968 was about 8.5 million tons (product weight). Peru was the leading exporter, accounting for about 28% of the total, and the United States was the leading importer, accounting for about 18% (Table 8.5). Other countries with larger shares of the trade, either exports or imports, were several in Western Europe, Japan, South Africa, U.S.S.R., Chile, Poland, and Morocco.

TABLE 8.5

Countries Importing or Exporting More Than 100,000 tons of
Fishery Products in 1968[a]

Importers		Exporters	
Country	Thousands of tons	Country	Thousands of tons
United States	1,546	Peru	2412
West Germany	982	Norway	819
United Kingdom	966	Denmark	497
Netherlands	426	South Africa	481
France	358	Japan	479
Japan	336	Canada	394
Italy	306	U.S.S.R.	320
Belgium	218	Iceland	287
Denmark	192	Chile	213
Sweden	186	Netherlands	199
Spain	171	Sweden	192
Poland	121	West Germany	113
		Morocco	104

[a]Source: Food and Agriculture Organization of the United Nations (1969). "Yearbook of Fishery Statistics," Vol. 27. FAO, Rome, Italy.

8.6 Recreational Fishing

Most people have strong emotions about the outdoors. They recognize that animals and plants form a web of life that includes man. They recall the struggle by them or their ancestors for survival and the feats of pioneers in exploring the wilderness or the sea. They feel, moreover, an attachment to the outdoors as something to be cherished and as a place to go for recreation.

Any examination of the recreational aspects of fishing is difficult because recreation itself eludes a simple satisfactory definition.* One man's work may be another man's recreation. It is easy to say recreation is not work, but neither are some religions or family activities. It is not satisfactorily defined as fun, or an unsystematic activity, or an unproductive activity. It is broadly defined as a change in social activity for one person from the usual family and work social groups toward a different group than the family group and a

---

*These concepts of recreation have been taken largely from: S. Z. Klausner (1969). "Recreation as Social Action, in a Program for Outdoor Recreation Research. Nat. Acad. Sci.— Nat. Res. Counc., Washington, D.C.

more homogeneous group than the work group. It can be considered as a dramatic function that binds the recreational group.

The form of recreation is highly varied. It may be sedentary or nomadic, in broad open spaces, such as the seashore, or in confined spaces, such as the forest. It may lead a person toward isolation or toward a group, and toward more vigorous or more passive activities. Much recreation involves the outdoors; walking, hiking, participatory games, such as tennis or golf, spectator games, such as baseball or football, camping, boating, or fishing. Much recreation depends on water, which is either essential to activities, such as swimming, boating, or fishing, or contributes to the enjoyment of activities at the camp on the lake or in the hotel at the seashore. All outdoor recreation is enhanced by clean surroundings that lack the imprint of thoughtless or selfish people.

Fishing caters to diverse recreational needs. It can involve a bus trip to a nearby pier or a transoceanic expedition. It can satisfy a solitary fly fisherman or a group on a party boat. It can be a lazy outing at a pond on a summer day or be an arduous search for trophies in a distant sea. It usually involves a trip, the outdoors, and associated activities such as boating, hiking or camping. It also involves uncertainties in the number and size of catch; a little gambling. Many people also enjoy fish watching. They may keep pet fish in aquaria (see Section 8.7.3 on culture of ornamental fish). They may also visit large public aquaria, fish hatcheries, fish passage facilities, or fish spawning areas.

2.6.1 METHODS OF RECREATIONAL FISHING

Almost all recreational fishing is angling conducted with hook and line; although occasionally it involves the use of dip nets, bow and arrow, spears, shovels (for clams), or other active gear operated by one or two persons. The concept of sportsmanship requires that the fisherman pit his individual skill against the fish and give the fish a chance to elude the gear or to struggle free. Large gear that requires two or more fishermen, fixed nets, or any really efficient gear is foreign to the concept.

Great efforts are made to catch trophy fish under handicaps. Artificial lures may be substituted for live bait, single barbless hooks for arrays of barbed hooks, weak lines and rods for strong ones. On the other hand, great ingenuity is exercised in devising the most attractive but artificial lure or most sensitive reel for handling the light line. The angler will also take pains to fish as comfortably as possible using special clothing, special boats, or special accommodations.

8.6.2 ROLE OF RECREATIONAL FISHING

A relatively large and increasing proportion of the people in the more economically advanced and urbanized countries turn to fishing for recreation. Many Europeans, Japanese, and North Americans engage in it wherever they may be in the world, and these ethnic groups are rapidly being joined by others. In the United States, fishing license sales are increasing at an annual rate of about 4%.

Good statistics on the practice of recreational fishing are available from the United States as a result of special surveys, the most recent of which was conducted for the calendar year 1965 (see U.S. Fish and Wildlife Service, 1966; also Deuel and Clark, 1968). About 28 million people, 12 years old or older (20% of the population), fished 3 days or more; about 45 million in all fished at least once. Those who fished 3 days or more spent $2.9 billion and traveled 22 billion miles. Of their expenditures 56% were for trip expenses, 37% for equipment, and only about 2% for license fees.

The participants came from all parts of the country, from broad spectra of occupations, residential locations, and income groups. The most noteworthy differences were that there were about twice as many men as women, a higher proportion of anglers among 12–15 year olds than among other age groups, and a higher proportion among people living outside metropolitan areas than within.

Saltwater fishermen made up a large fraction of the total. About 8 million anglers caught 740 million fish with a total weight of about 670 thousand tons. Angling was pursued all around the coast but intensively near New York City and Los Angeles.

The social consequences of an activity as popular as recreational fishing (Fig. 8.17) range from large economic effects to important political actions. Anglers who were seeking marlin off Mexico have even become involved in international incidents when they destroyed commercial fishing gear belonging to fishermen from other countries. The money spent by anglers supports travel media, fishing resorts, tackle businesses, and equipment businesses. The anglers demand and get public services such as improved roads, boat-launching facilities, and permanent public ownership of streams and lakes. The total catch of fish by recreational fishermen from salt water alone was nearly 50% of the total commercial catch in the United States for human food and, although no good statistics were presented for the freshwater catch, its addition would probably bring the total catch from recreational angling to a level well above the total commercial catch for food. Still another contrast appears in a comparison of the number of recreational fishermen with the number of commercial fishermen (about 129,000 in 1965). There were about 350 anglers for every commercial fisherman; a ratio that indicates the much greater political influence of the recreational fisherman.

Fig. 8.17. Angling in the United States is popular with men and women, young and old. (Photo courtesy of Washington Department of Game.)

8.7 Private Aquaculture

Private aquaculture supplies food, bait, recreational angling, and ornamental fish at a profit to the producers. It is like agriculture in that it requires nearly complete control of the animals or plants throughout their lives. It is unlike public aquaculture by government in that in the latter a profitable operation is not always necessary and usually the animals are released into a natural or only partially controlled environment.

8.7.1 AQUACULTURE FOR FOOD

Unfortunately, the statistics on aquacultural production are not generally available separately from the statistics on all fishery production. The best available estimate of total annual fish production is, 2,620,000 tons (Table 8.6). If we add the total production of oysters and mussels (Tables 8.7, 8.8), little of which is produced from uncultivated stocks, the total animal produc-

TABLE 8.6

COUNTRIES ESTIMATED TO PRODUCE ANNUALLY MORE
THAN 10,000 TONS OF FISH BY AQUACULTURE[a, b]

Country	Thousand metric tons
Peoples Republic of China	1190
India	480
U.S.S.R.	190
Indonesia	141
Thailand	136
Philippines	87
Japan	85
Taiwan	59
United States	40
Pakistan	38
Malaysia	26
Hungary	22
Viet Nam	13
Rumania	12
Poland	11
Czechoslovakia	11
Israel	10
Total (including countries other than above)	2620

[a]Source: T. V. R. Pillay (1970). Problems and priorities in aquaculture development. MS.; and personal communication from H. Kasahara for Japanese data (1970).

[b]Data are most recent available—not for a single year.

TABLE 8.7

COUNTRIES REPORTING A PRODUCTION OF
MORE THAN 10,000 TONS OF OYSTERS[a]
IN 1968[b]

Country	Thousand metric tons
United States	389
Japan	267
France	45
Korea	38
Mexico	36
New Zealand (1967)	13
Total world	823

[a]Unknown, but probably small fractions of the quantities are wild oysters.

[b]Source: Food and Agriculture Organization of the United Nations. (1968). "Yearbook of Fishery Statistics," Vol. 26. FAO, Rome, Italy.

TABLE 8.8
COUNTRIES REPORTING A PRODUCTION OF
MORE THAN 10,000 TONS OF
MUSSELS[a] IN 1968[b]

Country	Thousand metric tons
Netherlands	114
Spain	69
France	28
Chile	23
Dnemark	14
Italy	14
Federal Republic of Germany	11
Total world	305

[a]Unknown but probably small fractions of the quantities are wild mussels.

[b]Source: Food and Agriculture Organization of the United Nations. (1968). "Yearbook of Fishery Statistics," Vol. 26. FAO, Rome, Italy.

tion from aquaculture was about 3,753,000 tons live weight in 1968. Plant production in 1968 from intensive aquaculture was about 258,000 tons, mostly from eastern Asia. (The total aquatic plant production from the world in 1968 was 890,000 tons, and much of it was produced by methods of semicultivation, hence an accurate estimate of plant production from aquaculture is not possible.) Thus the total live weight of fishery food products from aquaculture was about 4,000,000 tons or 6.3% of the world total fishery production.

Almost all of the freshwater animal production was of fish, principally the several species of carps that are cultivated very widely in Asia and Europe. Most of the remainder consisted of catfish, trout (Fig. 8.18), eels, and tilapia. Almost all the saltwater animal production was of mollusks, principally oysters and mussels, and the rest largely from milkfish and yellowtail. Many other species of fish, mollusks, and crustaceans are cultivated in small quantities for food, and some (e.g., shrimp) are the object of intensive research, but the total production from such minor species is very small at the time of this writing.

Much of the recent interest in aquaculture has arisen from the large possible increase in yield of animals per unit of area. The potential annual yield from natural waters ranges from about 10 kg/ha for the most barren to a maximum of about 200 kg/ha for the most fertile. The yield can be increased to about 500 kg/ha through control of the populations of both predators and

Fig. 8.18. An aerial view of the Snake River Trout Co. at Buhl, Idaho. This "farm" can raise trout at a rate of about 400,000 kg/ha/year. (Photo courtesy of Snake River Trout Co.)

competitors and by encouragement of those animals that are low in the food chain (e.g., milkfish). Production can be increased to about 5000 kg/ha/year by the addition of food that can be consumed directly by the animals, but when yield reaches this level, the still waters in ponds are near the limit of their capacity to recycle the animal wastes. Greater production can be realized only by the addition of more food and augmentation of the flow of water to supply oxygen and carry away wastes. When both are done optimally with an especially tolerant species, the yield may increase enormously—to a maximum of about 2,000,000 kg/ha/year.

The potential yield from the cultivation of mollusks follows these principles although they feed entirely on natural foods. They are placed on beds or racks or suspended from rafts in flowing natural waters that bring suitable foods to them. Under some conditions the yield of mollusks has reached 300,000 kg/ha/year, but the food utilized is produced by a much greater area of water than that occupied by the mollusks.

The basic principles of operating any private aquacultural enterprise are determined by the degree of legal control over the enterprise and by economic considerations, namely, the market for the product, the costs of overcoming the technical problems, and the investment required.

The degree of legal control is, of course, complete if the water supply and land is wholly owned, but such control is rarely to be had. Usually the water must come from a public source and be secured by a water right. The land on which the operation is located may be wholly owned or leased from government if it underlies a public waterway or is part of an estuarine system. In such cases, the government agency needs to protect the public interest and commonly does so by rigid restrictions on the kinds of activity permitted, by fees based on production, and by short-term leases.

The market for fish tends to be taken for granted by those who see the protein deficiencies around the world, but most aquaculture produces high-cost items that supply luxury markets. The exceptions are the small farm pond operations in which the cost of production is combined with costs of other farm operations. An example that indicates how the cost of aquaculture compares with the cost of producing fish from the sea is provided by data from the Japanese*: the production of 46,000 tons of fish from aquaculture in 1965 required about 241,000 tons of fish for feed and about 40,000 tons of other foodstuffs.

The costs of operations are determined by many factors, some of which are influenced greatly by the choice of site. A water supply that is available in a gravity flow is cheaper than pumped water. A water supply that contains no fish eliminates a constant source of infectious diseases. An installation that is completely under legal control, i.e., in nonpublic water, is cheaper to operate than one inhibited by changing government rules. Food is the major cost item of intensive aquaculture; it must provide an adequate diet and be as cheap as possible. Labor is the next largest cost item, and the labor required to operate and maintain the installation is a necessary consideration in its planning.

Because of the high costs and considerable risks to be expected in aquaculture, the investment opportunities will rarely be as attractive as those for agriculture or industry, but must be sought where aquaculture has some special advantage. These may be found on lands too wet or too acid for agriculture, with water too salty to reuse for irrigation, or too warm to release into the natural environment, and where waste organic material suitable for food is available, and where the production can be marketed at an attractive price. When such a situation exists it may be possible to find a species suited to the water and markets and a mode of operation that can be profitable. It may be an extensive "ranch" type operation producing only a small yield per unit of area or an intensive "feed lot" type operation

*Personal communications from Hiroshi Kasahara (1970).

producing a large yield per unit of area. The maximum profits are not necessarily associated with the more intensive operations. It should be noted that giving priority to a good place for aquaculture is the reverse of that used by some people who think first of a choice species (e.g., lobster, pompano, sole) that could be raised for food. The costs of production may be entirely too high for the operation to be profitable after all of the technical problems are controlled.

8.7.2 AQUACULTURE FOR RECREATION

Private aquaculture for recreation is similar to and, indeed, a development from aquaculture for food. It consists of producing bait for anglers and operating private angling ponds in the United States, where apparently it is best developed.

Most of the fish grown for bait are minnows of the family Cyprinidae; the principal species are the golden shiner, *Notemigonus chrysoleucas*; the fathead minnow, *Pimephales promelas*; and the goldfish, *Carassius auratus*. All are adapted to warmwater pond culture and are grown in substantial quantities in the states bordering the southern Mississippi River. No overall statistics on production are available, but it is probable that several thousand hectares of ponds are used for intensive culture of bait fish.

A large number of people who enjoy angling in the United States are catered to by aquaculturists who provide fishing for a fee, usually to catch trout. The ponds, called "catch-out" ponds, are kept heavily stocked with fish. Such recreation is becoming increasingly popular near centers of population where natural waters cannot fill the demands of the anglers.

Another more casual form of aquaculture is practiced in farm ponds used primarily to supply water to livestock. A supply of fish for food or fun is a common secondary use in many countries. In the United States there were an estimated 900,000 ha of such ponds in 1965.

8.7.3 ORNAMENTAL FISH

Fish were kept as pets many centuries ago, according to scattered reports in the literature, and the practice probably was started long before it was recorded. The people of eastern Asia first developed varieties of goldfish to keep in garden ponds, but interest in keeping ornamental fish did not spread to countries of the west until about 1850. This development followed the discovery that fish require oxygen from the air or from plants if they are to be kept in an aquarium. Soon after 1850, public aquaria opened in many cities in Europe and goldfish were introduced into America from Japan. The hobby included at first other northern species of fish, but after 1900 an increasing number of tropical species became available. Now, several hundred species of tropical fish are available through dealers in many parts of the world.

The hobby of aquarium-keeping is fostered by many related activities. Aquarium societies have been formed in many cities. Aquarium journals and books about aquarium keeping are published in many languages and seem always to attract the spectacular efforts of color photographers and color engravers. Tanks, stands, filters, aeraters, heaters, and other equipment are furnished as well as food and chemicals for disease treatment by aquarium supply stores.

Aquarium keeping can be a fascinating exercise in aquatic ecology. An aquarium is commonly arranged with a bottom of sand and supplied with a variety of plants and animals. The plants are chosen for their decorative effects. The animals usually include colorful fish and scavengers such as catfish or snails. The tank is heated or cooled to a temperature suitable for the species chosen and the water aerated and filtered as necessary.

The plants and animals supply some of each other's needs, but an aquarium can never be a balanced ecosystem without a certain amount of maintenance. The animals produce CO_2 and wastes that are utilized by the plants, and the plants produce O_2 and some food for the animals, but the balance must be maintained by imputs. Oxygen is supplied and excess CO_2 removed by aeration. Excess waste materials and plankton are removed from the water by filtration. Food is supplied in amounts and kinds needed by the principal animals.

The fish are available in a great variety of colors, shapes, and habits. Many exhibit distinctive courtship and breeding behavior; some bear their young alive, others lay eggs. Many go through elaborate behavior in caring for their young; others devour them. The color and behavioral variations among the species available are almost endless and limited only by the budget and time of the aquariist.

The economic importance of this hobby is not well documented, but scattered information indicates that the fish sales alone may be a considerable fraction of the fish business in the United States. It has been asserted that 22,000,000 households (about 40% of the total) keep aquaria (Axelrod, 1969). Importers declared 73,300,000 live fish to United States Customs in 1969. Such imports are only a small fraction of the estimated total United States supply, which comes mostly from private fish farms. The majority (about 350) of these are located in Florida, from which an estimated 200,000,000 fish were shipped in 1969.

In addition, the equipment, supplies, and publications sold to aquariists are probably much more valuable than the fish. Note, for example, that brine shrimp eggs, which are used mostly for aquarium fish food, have been estimated to have a wholesale value to the United States producers of about $1,000,000 annually.*

8.8 Public Aquaculture

As the wild stocks of aquatic animals have become depleted people have tried to control the natural environment for the benefit of the animals and to augment the stocks by aquaculture. The wild animals are public property, and the efforts to assist them have been mostly governmental rather than private. Most of the public interest in the animals is recreational rather than economic. Such public aquaculture has developed most extensively in the United States and Canada. Public aquaculture to assist private aquaculture through supplying eggs or brood stock is practiced in some countries but will not be discussed further here.

Public aquaculture started in France in 1852 when the first public trout hatchery was constructed. Its construction followed the discovery and rediscovery during the preceding century in Germany, England, and France of methods for the manual fertilization of trout eggs, the care of the eggs during incubation, the care and feeding of young trout, and the transportation of both eggs and young fish. These techniques made possible the economical production and transportation of large numbers of eggs and young. The promoters of the first hatchery argued that they could restock all of the streams in France at a negligible cost to the government. They suggested that adoption of their techniques would increase the annual revenue from the waters of France from 6,000,000 francs to more than 900,000,000 francs.

News of the claims for the efficacy of these "new" methods soon reached the United States, where there was much public concern about the depletion of the natural waters. Many of the Eastern States formed fish and game commissions during the 1860's, and the federal government formed the United States Fish Commission in 1871. The first United States trout hatchery was built at Mumford, New York, in 1864. Many others were soon built, some by government and some by private operators who sold trout and eggs to the new government hatcheries. Some were built for other species, such as a Pacific salmon hatchery at McCloud River, California, in 1873, and a marine fish hatchery at Woods Hole, Massachusetts, soon afterward.

*R. R. Whitney (1967). Introduction of commercially important species into inland mineral waters, a review. *Contrib. Mar. Sci.* **12**, 262–280.

The hatcheries seemed the means for the conservation of fish. Foreign scientists had made extravagant claims for their benefits, using arithmetic that was most convincing. Each trout produces 1000–2000 eggs, most of which can be hatched into fish and placed in the natural waters to grow. It costs very little to do so, and it was entirely believable that the fish population could be augmented a thousandfold. Politicians soon discovered that building hatcheries helped them to get elected, and thus the operation of government fish hatcheries became a major part of the fish conservation movement in the United States. The number of public hatcheries grew to about 68 in 1900, half of them operated by the federal government and half by the states, and to 629 in 1948. Subsequently, some of the less efficient hatcheries were closed, and the number declined to 502 in 1965, 92 of which were operated by the federal government and 410 by states.

TABLE 8.9

FISH PRODUCED IN THE UNITED STATES BY FEDERAL AND STATE HATCHERIES (1965)[a]

	Millions of			Thousands of metric tons
	Fry	Fingerlings	Catchables and yearlings	
Salmon and steelhead	54	216	17	1.4
Trout and grayling	61	124	64	8.1
Catfish	6	17	1	0.2
Other warmwater fish	1045	119	1	0.4
	1165	476	83	10.1

[a]Source: U.S. Fish and Wildlife Service, 1968. National survey of needs for hatchery fish. *Bur. Sport Fish Wildl. (U.S.), Res. Rep.* **63**, 1–71.

The production (by weight) from these hatcheries in 1965 (Table 8.9) was predominantly of trout. The total of 8100 metric tons represented an average of 1.1 kg per coldwater fisherman in the United States. About 91%, by weight, of the trout were of catchable sizes; a much greater proportion than in earlier years, when most were released as fry or fingerlings.

The production of fish of warmwater species other than catfish consisted mostly of largemouth bass, bluegill, redear sunfish, northern pike, muskellunge, walleye, and striped bass. These and fish of numerous other species were distributed largely as fry because of the high cost of feeding and rearing to fingerling or larger sizes. No satisfactory formulated feeds have been found for most of these species, and fresh raw meat or fish is used. Most species are also cannibalistic; thus survival is low to fingerling or catchable sizes.

Despite the rapid growth of the fish hatchery program in the United

States from 1870 to 1940, the quality of angling was not satisfactory. Special ecological studies of lakes and streams that started during the 1930's led to the surprising conclusion that poor fishing in warmwater lakes was frequently due to too many fish! These were either fish of undesirable species that inhibited those of game species or stunted fish of desirable species. Other studies revealed that water quality was unsuitable for some species but adequate for others and that certain combinations of species provided much better fishing. Such findings led to an intensive search for methods of population control that could be applied to large natural bodies of water and to studies of methods of habitat improvement. All such findings pointed out clearly that aquaculture and the stocking of fish were only two of numerous techniques that needed to be considered.

The demand for more control of recreational fishing waters became more urgent after World War II as large areas of water were impounded as reservoirs for irrigation, municipalities, power development, levee construction, recreation, or as lakes filling basins excavated for mines or gravel pits. All of these became potential fishing sites that would provide better fishing with suitable construction and management than without.

The consequence of such needs and demands has been the development of an extensive (as contrasted to intensive) public aquaculture in a large proportion of the waters in the United States. The aquatic environment is usually either the natural environment or a man-made body of water constructed for some purpose other than fishing, yet a considerable degree of control can be exercised over its quality. Temperature can be controlled to some extent by proper location of the depth of the outlet from a reservoir. Dissolved oxygen can be supplied occasionally by mechanical means or by controlled circulation in a reservoir. Fertility can be augmented or decreased. Pollution can be reduced. Fish shelters or fish spawning areas can be provided.

In addition to the environmental controls there are some practical population controls. Complete fish populations can be removed by poisoning and desirable species introduced. Reservoir levels can be changed to inhibit the spawning of undesirable species. The population of undesirable species can be reduced by selective poisoning and netting. Disease-resistant strains can be introduced. If these measures do not bring the fish production up to satisfactory levels, then the waters can be stocked with catchable-sized fish.

The exercise of such environmental and population controls is commonly done with the least possible interference with other uses of the water. It requires extensive coordination of the fishery agency activities with those of other agencies and individuals concerned with the water and, together with the exercise of regulatory authority, comprises the principal inland fishery management activity in the United States.

REFERENCES

The kind of information available on the various fisheries is relatively uneven. The food fisheries on the public resources are largely well documented with respect to history, methods, and amount of catch. The recreational fisheries have stimulated an immense literature in many languages about the history, methods, and pleasures, but the extent of the fisheries and their catch is poorly documented. Even less information is available about aquaculture; the actual practices of private food fish producers are proprietary to some extent and the data on total production are mixed with the data on production from wild fish. The operations of the producers of ornamental fish are almost completely unrecorded.

Alverson, D. L., and Lusz, L. (1967). Electronic role in the fishing industry—present and future. *Proc. IEEE, Int. Conv. Rec.* Part 8, pp. 3–11.

Axelrod, H. R. (1969). "Tropical Fish as a Hobby," rev. ed. McGraw-Hill, New York.

Borgstrom, G. (1961). "Fish as Food," Vol. 1 Academic Press, New York.

Borgstrom, G. (1962). "Fish as Food," Vol. 2 Academic Press, New York.

Borgstrom, G. (1965). "Fish as Food," Vol. 3, Part I, and Vol. 4, Part II. Academic Press, New York.

Deuel, D. G., and Clark, J. R. (1968). The 1965 salt-water angling survey. *Bur. Sport Fish. Wildl.* (U.S.), *Res. Rep.* **67**, 1–51.

Food and Agriculture Organization of the United Nations. (1959). Modern fishing gear of the world. *1st FAO World Fishing Gear Congr. 1957* Vol. 1.

Food and Agriculture Organization of the United States. (1964). Modern fishing gear of the world. *2nd FAO World Fishing Gear Congr. 1963* Vol. 2.

Food and Agriculture Organization of the United Nations. (1969). "Yearbook of Fishery Statistics," Vol. 27 (see also earlier volumes). FAO, Rome, Italy.

Fry, W. H. (1854). "A Complete Treatise on Artificial Fish-Breeding." Appleton, New York.

Gabrielson, I. N., and LaMonte, F., eds. (1963). "The Fisherman's Encyclopedia," 2nd rev. ed. Slackpole & Heck. This is a major source of information on recreational fishing.

Heen, E., and Kreuzer, R., eds. (1962). "Fish in Nutrition." Fishing News (Books) Ltd., London.

Innes, W. T. (1949). "Goldfish Varieties and Water Gardens," 2nd ed. Innes Publ. Co., Philadelphia, Pennsylvania.

Innes, W. T. (1966). "Exotic Aquarium Fishes," 19th rev. ed. Metaframe Corp., Maywood, New Jersey.

Iversen, E. S. (1968). "Farming the Edge of the Sea." Fishing News (Books) Ltd., London.

Kreuzer, R., ed. (1965). "The Technology of Fish Utilization." Fishing News (Books) Ltd., London.

Pax, F., ed. (1962). "Meeres produkte—ein handworterbuck der marinen rohstoffe." Borntraeger, Berlin.

Radcliffe, W. (1926). "Fishing from the Earliest Times," 2nd ed. Dutton, New York.

Stansby, M. E. (1963). "Industrial Fishery Technology," Von Nostrand-Reinhold, Princeton, New Jersey.

Traung, J.-O., ed. (1955). Fishing boats of the world. "International Fishing Boat Congress, 1953," Vol. 1. Fishing News (Books) Ltd., London.

Traung, J.-O., ed. (1960). Fishing boats of the world. "International Fishing Boat Congress, 1959," Vol. 2. Fishing News (Books) Ltd., London.

U.S. Fish and Wildlife Service. (1966). 1965 survey of fishing and hunting. *Bur. Sport Fish. Wildl.* (*U.S.*), *Res. Rep.* **27**, 1–76.

U.S. Fish and Wildlife Service. (1968). National survey of needs for hatchery fish. *Bur. Sport Fish Wildl. (U.S.), Res. Rep.* **63**, 1–71.

Von Brandt, A. (1964). "Fish Catching Methods of the World." Fishing News (Books) Ltd., London.

Fishery Resource Management

People have long recognized their common interests in the fisheries, but because the animals are usually regarded as belonging to no one until caught or until produced by aquaculture in confinement, their care and protection has been neglected. The fishery resources can be perpetually available if they are suitably protected and improved. Due to the failure to provide for the protection and improvement of the resources during the course of normal political bargaining, the need for sound technical information and for government decisions based on technical facts has been realized. Fishery agencies have been organized in the administrative branches of most governments to meet this need.

9.1 Functions of Fishery Resource Agencies

Fishery resource agencies are established by either administrative or legislative action that delegates to them certain authorities and responsibilities. Usually the action recognizes the public ownership of the resources and the need for an informed and effective agency to manage the resources in the public interest. The agency is organized to provide basic information on the resource, to employ trained people familiar with the problems, and to supply an organization capable of acting on the problems. With the ability to act and a history of having done so effectively, the agency can gain the confidence of the public and the support it needs. It can develop specific programs within its mandate, which are usually stated in very general terms. Therefore, it is appropriate to examine first what fishery resource agencies actually do rather than to examine the authorities and responsibilities delegated to them. Their major functions are:

1. *Collect and evaluate facts concerning the resources.* Commonly, resource agencies are intended by the administration or legislatures that formed them to operate on the basis of facts and not merely on the basis of political pressures. The agencies need basic statistics on the use of the resource, such as catch, effort, participants, supplies, and markets. The agencies also conduct special investigations of problems ranging from short-term studies, for the purpose of immediate decisions, to long-term basic research for the purpose of better understanding.

2. *Provide public information.* The basic statistics and the results of the investigations are commonly disseminated through publications, press releases, and special presentations, either to legislatures or to public groups. Some agencies also conduct organized educational programs.

3. *Protect and improve the aquatic environment.* Many departments have authority to review plans for structures, such as dams, bridges, and water diversions that may affect the aquatic life. They also have the authority to establish water quality standards and to control pollution, or they may do so through collaboration with other agencies.

4. *Provide public facilities for the use of the resources.* This function may include the design and construction of port facilities or special means of access to inland fisheries.

5. *Establish regulations as permitted by legislative bodies.* These may be designed to conserve the resources, prevent waste, divide the resources among these who want to fish, and protect consumers. These functions are commonly based on the results of special studies by the agency as well as information obtained through public hearings.

6. *Enforce the legislated laws and department regulations.* Generally, this activity requires a large proportion of the agency budget and personnel. When the agency has responsibility for high seas fisheries, it may be especially difficult and expensive to perform.

7. *Propagate, distribute, and salvage fish.* In North America these functions are largely concerned with recreational fisheries, but in many other countries they may be concerned also with commercial aquaculture.

8. *Advocate the cause of the aquatic resources.* In many circumstances the agency is the best source of information concerning the fishery resources and plays the role of their defender against possible damage by pollution or use of water for other purposes. It may also be an advocate of the political causes of the fishermen even though this function is not a delegated nor proper function for an administrative agency.

Other important functions occasionally assigned to fishery agencies are not concerned with managing the resources. One is the provision of social services to commercial fishermen. These may include the medical services that are provided to seamen (including fishermen) in many countries of the world. Agencies may also supply educational services either to fishermen or to their families, or to their villages and economic assistance such as supplemental payments on catches or subsidies on equipment or low-interest loans. Some forms of such special assistance are given to commercial fishermen in most countries in recognition of their special needs and their extraordinary valor in following a seagoing life.

Another function is the setting and enforcement of sanitary regulations designed to protect the public health. Many aquatic foods are unwholesome when handled inappropriately, and some, especially certain mollusks and crustaceans, are unwholesome at certain seasons; regulations are therefore essential in the public interest. Regulation is commonly combined with an inspection program, the findings from which furnish the basis for action under the regulations.

9.2 Origins of Public Policy

The present functioning of fishery resource agencies develops basically from public attitudes that have evolved for a very long time. It must be presumed that the earliest people who subsisted largely by hunting and fishing originated the concepts that the game and fish belong to no one until after they are killed and physically possessed. Such wild animals were important to everyone who depended upon them, yet with a few exceptions, they could not be restrained and possessed while alive. If someone with unusual ingenuity managed to possess or claim the wild animals, he would have been considered to be reducing the resources for other people.

Apparently such natural law with respect to fish has been followed through the ages. In early Roman law, fish were placed in the category *res nullius*, things that belonged to no one, on the basis of the argument that this classification was in conformity with natural or moral law. Similar principles obtained by law in medieval times in Europe continue to the present. In most countries fishing laws are still based on the assumption that the fish belong* to no one until caught, except for certain government waters and for people practicing aquaculture.

Since fish have commonly been considered to belong to no one, fishing has been regarded by everyone as a right. In ancient China, all waters were

---

*They are sometimes designated as common or public property, but they are not owned in the sense that most property is.

free and open to fishing, with the exception of a few imperial preserves. Under Roman Law, the sea and public waters were open for fishing by anyone. However, private waters were recognized and granted by governments in such situations as coves, backwaters, small lakes, and aquacultural ponds. Exceptions were also made in some places in favor of the fisherman who first occupied a site with fixed gear, such as a trap. He was allowed exclusive use of the shore or water for a reasonable distance around his gear even though he did not own the fish until he caught them. Today, most domestic waters remain open to fishing by all citizens of the country having jurisdiction. In countries with large numbers of recreational fishermen, this right may be exercised by a large proportion of the population.

Those who fish in international waters, namely the sea and a few large boundary lakes or rivers, go beyond the limits of domestic authority into an area that, in early times, was shared universally. It was an area of no law, an area of freedom from all domestic authority. Freedom of the seas remains a fundamental concept in their use by fishermen and others, but its history since about 1300 A.D. is one of gradual reduction.

As states have developed their central authority over domestic affairs, many of them have asserted authority over nearby waters. These waters have ranged in extent from semienclosed bays and harbors to large sections of the sea. The most extensive were probably the claims that resulted from the papal allocation of large parts of the world oceans to Spain and Portugal in the fifteenth and sixteenth centuries. These authorities included various rights, such as those of passage and fishing, even those of ownership and collection of taxes.

Such assertions of extended authority were challenged vigorously in 1609 by Hugo Grotius, a Dutch lawyer. He argued that what cannot be seized or occupied cannot become property, that the sea is free to all, and fishing is free to all men. He argued further that there was not moral justification for extended authority over fishing because the sea cannot be exhausted by promiscuous use. These arguments were advanced by the Dutch to support their interest in fishing for herring off the English and Scottish coasts. They were soon countered by British assertions of authority over coastal waters and arguments that the authority was essential for conservation. The disagreement gradually deteriorated into questions of sovereignty, the honor of the flag, and the series of Anglo-Dutch wars.

In the following century (eighteenth) considerable agreement developed regarding the concept of freedom of the seas as well as the need for authority over a limited territorial sea in coastal waters. The authority was needed for defense, among other reasons, and it was argued, therefore, that the width of the territorial sea ought to be the distance a cannon could shoot. Eventually (in the early twentieth century) 3 miles was endorsed as the breadth of the territorial sea by most of the maritime nations. There continues, however, a widening of its extent, first, as a consequence of disputes over fishing in certain bays or along irregular coastlines. Larger and larger areas have been included, either through mutual agreement or world court decisions, in the internal waters of coastal countries or their territorial seas (Fig. 9.1), on the basis of the claim that baselines should be drawn from headland to headland. At first such baselines could be no more than 10 miles, but recently longer baselines across internal waters have been recognized. Second, a few countries have claimed authority over much wider territorial seas, notably 200 miles by several countries of South America.

The arguments over fisheries throughout these centuries were primarily over the authority to exploit rather than the authority to conserve, even though the need to conserve was frequently cited. The British wanted to exclude the Dutch from fishing along their coast on the basis of the argument that their fisheries were subject to exhaustion from overfishing and claimed the authority to conserve them. The British were, of course, quick to blame

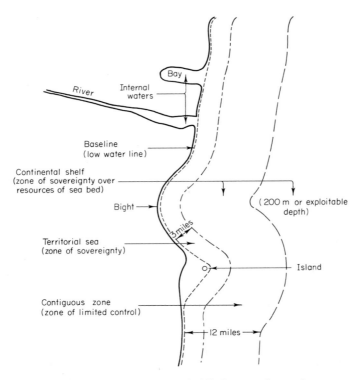

Fig. 9.1. Diagram of the common juridical status of coastal waters.

any declines in the stocks on the foreign fishermen, which included in later years the French and Danish, as well as the Dutch. The British were not, however, hypocrites about conservation with regard to other countries. They passed numerous domestic laws, starting in the fourteenth century, for the protection of migrating salmon, for the protection of young fish, and for the protection of spawning fish. The need for conservation, although hotly debated and illy defined, remained a widely held belief of the British people which the colonists took with them to North America and here it has remained deeply rooted. Such a belief in conservation has not been shared by many other countries, and no significant international agreements for fish conservation were made until after the beginning of the twentieth century.

9.3 Objectives and Methods of Management

Long before we acquired any scientific understanding of populations, fishermen became concerned when they discovered that they could reduce the abundance of the stocks of wild fish by overfishing. Recognizing that individual restraint would not work unless practiced by all, they tried to restrain the fishing by laws. They were acquainted with the needs of domestic animals; they considered it prudent to prevent waste by avoiding the use of animals not in prime condition, to avoid interfering with reproduction by protecting animals, especially females, during the breeding season, and by protecting the young to ensure their growth. Hence, laws were promulgated for these purposes.

When the abundance of a stock still declined after these measures, the fishermen objected to fishing by outsiders in their fishery and to excessive ingenuity on the part of any fisherman. Thus an efficient new net would be objectionable to the nonusers because the stocks might not stand the fishing if everyone used it and because its use by a few would give them a special advantage. Consequently the fishermen pressed for laws to prohibit the use of efficient new gear.

When the abundance was still inadequate after the fishermen had been prudent and inefficient and successful in keeping the resource to themselves, the fishermen sought to place the blame on pollution, on other changes in the environment, or on adverse natural events. These could be alleviated in some instances but often were not, because the fishermen did not carry the political weight to battle the polluters or others who were depleting the resource. Consequently, when another ray of hope emerged in artificial propagation, the fishermen pressed for this as hard as they could.

Unfortunately the long history of restraints and other measures to protect the food fisheries has shown few benefits to the resource or to the fishermen, probably because most measures were too little and too late. This lack led to the denunciation of restraints, by the eminent scientist T. H. Huxley in 1883, among others, and has left a doubt in the minds of many about the usefulness of regulating the fisheries.

Regulations that have been long advocated for the protection of the stocks on grounds of prudence and that are still in widespread use in the food fisheries are of the following types:

1. *Closed seasons*. Fishing is prohibited either during the spawning season or during any season of poor-quality fish.

2. *Closed areas*. Fishing is prohibited in either spawning or nursery areas. This regulation may be used as an alternative to closed seasons when either the spawners or young congregate separately from other parts of the stock.

3. *Prohibited methods*. Wasteful or dangerous practices, such as the use of explosives for fishing, are usually prohibited. Unusually efficient methods are occasionally prohibited when they affect the welfare of large numbers of inefficient fishermen.

4. *Protected females*. Female lobsters and crabs that carry the eggs and young beneath the abdomen are commonly protected when carrying eggs. In some species in which the female is small and less desirable for harvest than the male, such as the Alaska king crab, the females are protected at all times. Female mammals may be protected when accompanied by young or at all times.

5. *Protected young*. A minimum size limit on fish kept may be required when young animals can be released unharmed or when concentrations of young animals can be avoided by fishermen. If the fishing gear prevents the fisherman from releasing the small animals alive, a minimum mesh size may be imposed instead that will allow small animals to escape before the gear is hauled. Such a measure is practical only with gear that catches mostly a single kind of animal. For example, most trawls catch several kinds of fish or different shapes and sizes and therefore retain undesirable small individuals of the large species as well as desirable individuals of the smaller species.

None or even all of these regulations are likely to provide maximum, sustainable, physical yield from a stock or even prevent serious overfishing (short of total closures). They have been applied for centuries and may indeed be helpful, but cannot assure the modern objectives of conservation. Other regulation is needed.

The regulation must include control over the proportion of the stock caught (see Chapter 6). Such control requires knowing the size of the stock and the effects of fishing. Control over catch may be accomplished in part by closures, protection of spawners or imposition of size limits, but can be realized fully only by the additional enforcement of a quota that is set and changed according to the measured effects of fishing on the stock. When a quota is used it is possible to achieve maximum sustained physical yield, but this accomplishment still does not ensure the welfare of the fishermen. Control of the amount of fishing is also necessary (discussed in Section 9.4).

Regulation of food fisheries is necessary also for purposes other than conservation, which can best be described as orderly fishing. A large number of regulations fall in this category. The enforcement process requires that fishermen and dealers be identified, i.e., licensed. The license fee may include a tax for the privilege of going fishing or taxes levied on the catch. Basic statistics on catch, method, location, etc., are frequently required by regulation. Enforcement procedures are specified in the regulations, including means of informing the public as well as the actions of officers. Penalties for violations are prescribed.

In addition, regulation of food fisheries may involve division of the catch among groups of fishermen. Different gear and vessels are used, and each will catch better than the others under different circumstances. When one kind of gear is favored over another by conservation regulation, as happens frequently, some fishermen may oppose the regulation on grounds of equity and force reconsideration. Further, it is common practice in many countries to provide protection for the poorly equipped, inshore fishermen against the fishermen with large efficient vessels by reservation of the use of nearby grounds to the inshore fishermen.

For recreational fisheries the objectives are not as simple as for food fisheries. The maximum sustainable yield may not be equivalent to maximum recreation since it may depend on fishing in pleasant surroundings, on catching a fish in especially challenging circumstances, on friendly competition for the fish, or on associated activities, such as boating, hiking, or camping. High on the priorities for enjoying recreational fishing may be exhibition of individual rather than group prowess, avoidance of efficient gear, and confidence that the resource is for the fishermen to enjoy. The catch may be important only as it relates to the circumstances and may either be released alive or be killed for consumption later. Catches that are made too easily may even be despised or discarded.

The regulation of recreational fisheries needs to fulfill as many of these varied objectives as possible, but especially, the division of the catch among fishermen as fairly as possible by the imposition of individual catch limits, gear restrictions, size limits, closed seasons, and closed areas. The division needs to be devised also with the objective of maintaining the resource at a level of abundance acceptable to the anglers.

Regulation of the fishing is not, however, the only method of management available to most fishery agencies. They may prevent deterioration of
the aquatic environment or undertake a limited application of aquacultural
methods for the maintenance or enhancement of the wild fish stocks through
authority delegated to them, collaboration with other agencies having
authority, or publicizing of information about the resources. They protect
the environment against deterioration primarily by controlling the competing uses of the water (a topic of Section 9.5).

Application of aquacultural methods to the wild stocks in their environment is practical only in estuaries or fresh waters where the environment
and the animals can be physically controlled, at least to a limited extent.
The methods available include all of the controls that are possible in intensive aquaculture, but they are useful in public waters only where the circumstances provide special benefits.

The most common practice, by far, in North America and a few other
countries is partial population control, primarily by the stocking of artificially reared young (see section 8.8 on public aquaculture). This control
includes also the elimination or reduction of the populations of unwanted
fish and the transfer of preferred wild species to new environments. It may
include the avoidance through regulation and inspection of the transfer to
natural waters of undesirable species, such as ornamental fish or bait fish.
It may also include the development and stocking of new varieties developed
either through hybridization or selective breeding.

The second major practice is environmental control (in addition to
defense of the environment, Section 9.5). It includes construction of public
fishing waters, fertilization, removal of barriers to migration, construction
of fish shelters, control of vegetation, control of temperature at dams, and
control of oxygen content.

The other common aquacultural controls on disease and nutrition are
rarely possible in public waters. Occasionally supplemental feeding with
artificial food or isolation of diseased fish may be done directly, but usually
either kind of assistance must be provided indirectly through population or
environmental control.

9.4 Economic Problems of Fishery Management

The widely adopted objective of managing the food fishery so as to obtain
the maximum sustainable yield of each stock has been sharply challenged by
numerous economists because of the economic troubles of many fishermen
in the older fisheries. They argue that the greatest physical yield does not
necessarily provide the greatest revenue, the greatest profit, or the greatest

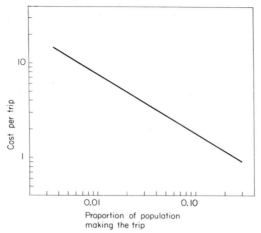

Fig. 9.4. A hypothetical demand function for participation in a recreational activity [After M. Clawson (1959). "Methods of Measuring the Demand for and Value of Outdoor Recreation," Reprint No. 10. Resources for the Future, Washington, D.C.]

made elsewhere. Second, a large proportion of the expenditures are not made in the area where the recreational opportunity exists.

The desirable base for evaluating recreational fishing would be the net economic benefits. This requires the estimation of a demand function (Fig. 9.4) showing what users would pay if the recreation were priced at different levels. The problem has been approached by the estimation of a demand function from data on cost related to the proportion of people using the recreational opportunity that can be obtained from people who travel different distances. Such a function can be assumed to apply to a particular recreational opportunity in an area with a given population.*

9.5 Conflicts with Other Uses of Water

Every fishery agency is involved or is likely to become involved in protection of the water used by the fish. This function may be as important as the regulation of the fisheries to the welfare of the fish resource; certainly it is so in estuaries or fresh waters and possibly in the sea. Man's ever increasing use of water and his contamination of natural waters pose serious problems for the fisheries.

*For details, see M. Clawson (1959). "Methods of Measuring the Demand for and Value of Outdoor Recreation," Reprint No. 10. Resources for the Future, Washington, D.C.; W.G. Brown, A. Singh, and E. N. Castle (1964). An economic evaluation of the Oregon salmon and steelhead sport fishery. *Oreg. Agr. Exp. Sta., Tech. Bull.* **78**, 1–47.

Any manipulation of a body of water must be expected to cause an ecological change in the life of the waters. Any ecological change must be expected to be beneficial to some organisms, harmful to others. Frequently the fishery manager will be the advocate of the fishery resource and will attempt to minimize the harm to the fisheries and to maximize the benefits. In order to do this, he must understand the water resources and their use.

Water is one of the more abundant substances on earth (see Section 2.1), but only a tiny fraction is available as fresh water in lakes and streams. This fraction of the circulating water is the runoff, or roughly one-third of the precipitation. This amount is still a large quantity relative to man's needs, and his problem is getting the water where and when he needs it in the quantity and quality required. A common characteristic of the supply in streams is a hundredfold to a thousandfold variation in the rate of flow during a year. Hence man tries to store large quantities in order to make it more uniformly available.

Water is legally regarded in most countries as a public resource not owned by anyone, but governments frequently assign rights to the use of the water.

The uses of water are commonly classified into withdrawal and flow, or on site uses. Withdrawal uses involve transport of the water, use, loss to the atmosphere, addition of dissolved or suspended materials or heat, and return to the natural system. The principal withdrawal uses are for municipal use, cooling of thermoelectric power plants, manufacturing, and irrigation. The principal losses through evaporation occur in irrigation, part of which is a municipal use, and in cooling of industrial or power plants. The flow or on site uses include hydroelectric power, navigation, recreation, provision of fish habitat, and waste dilution. The flow uses may happen concurrently with each other or with most withdrawal uses.

Water systems supplying water for withdrawal are usually installed by governments, which frequently charge a large fraction of the cost to the public rather than to the users. This practice may lead to a lavish use of water. The per capita consumption by municipalities in the United States is about 400 liters/day. Similarly, industries that get water very cheaply use processes or cooling methods requiring large quantities of water in preference to alternative methods.

Conflict between the needs of fish and other uses arises from withdrawal of needed water, from blockage of waters to migratory fish, from exposure of fish to hazards, and from modification of the quality of the water. The withdrawal from many streams can be greater than the low water flow, so the fishery agency may require minimum flows. Dams prevent migration, so the fishery agency may require fish passage facilities. Diversions to turbines or irrigation ditches may allow fish to go to their death, so the fishery agency may require screens.

The change in quality of the water with use is, however, the major hazard to fish in fresh water, estuaries, and even the sea. The rivers not only receive wastes directly from municipalities and industry but also receive the contaminants washed out of the atmosphere and those leached from the solid wastes deposited on the ground. All of the wastes go through the estuaries before final dilution by the sea.

For many years our principal concern over pollution has arisen from the hazard to health caused by human wastes. This hazard has been greatly reduced in most parts of the world by immunization against typhoid fever and by treatment of water supplies. Some hazards to human health remain but are much less important in most places than the effects of pollution on the use of water for other purposes.

The principal problem arises because we have long been accustomed to using the natural capacity of the waters to degrade our wastes but we have not kept within that capacity. Many materials, especially those of plant or animal origin, such as wastes from food-processing plants or sewage of human origin, are biologically degradable just as the wastes and dead bodies of the plants and animals that live in the water are degradable. The wastes from society are introduced into the natural waters in such quantities and in such limited areas that the capacity for degradation is exceeded. The consequences are lowering of the oxygen level, killing of the fish when the oxygen level is lowered enough, and creation of bad odors when all the oxygen is used.

The decomposed organic wastes along with the phosphates and nitrates contained in detergents and the drainage from agricultural lands also create excess fertility. The consequences are blooms of algae and other aquatic plants. As these plants become abundant they may choke the water by using all of the oxygen during the night which causes fish kills, or they may die and cause troublesome scums and odors. The process is called eutrophication, and it may ruin the value of the water to support fish or provide recreation.

On the other hand, some organic materials and inorganic wastes do not degrade rapidly or even at all. Some are toxic, expecially pesticides and compounds of heavy metals such as lead, mercury, copper, and zinc. Those that do not degrade accumulate in the environment and can become toxic eventually to fish or to people who eat the fish or use the water.

However, much greater dangers to fish or animals that eat the fish, including man, develop from the biological concentration of the nondegradable or slowly degradable substances. Many stable chemical elements or compounds are concentrated from the water by some aquatic organism. The concentration factors range up to tens of thousands, so that innocuous amounts in the water may become dangerous in the bodies of certain organisms. Further concentration may occur in the food chain when the materials are concentrated by predators as well as prey. This seems to happen especially with materials that are stored in skeletal or fatty tissue or are retained by the liver or spleen.

The potential of elemental concentration was discovered during efforts to trace the artificial radionuclides released around the earth by the nuclear explosions. The radioactive elements were traceable, and a few, notably strontium-90, were found to be biologically concentrated enough to present radiation hazards to man. The radioisotopes of the rare earths may present radiation hazard to aquatic organisms.

Elements that tend to concentrate in skeletal tissue may be hazardous in whole fish used for reduction to meal or fish protein concentrate. The fish are dried whole to about one-sixth of their wet weight, and all of the skeletal materials are concentrated. Fluorine and lead have been found in alarming concentrations in some samples of fish protein concentrate from marine fish.

Other slowly degradable substances that are concentrated alarmingly are the chlorinated hydrocarbon insecticides, notably DDT. This has a half-life in the soil of many years, so it accumulates during periodic use. It is also concentrated through the food chain of aquatic animals and is deposited in fatty tissue, including eggs. It has been shown to cause loss of fertility and, in some fish in central United States, to surpass the official concentrations permitted in food products.

With continuing pollution of the inland waters and the great temptation to use the sea increasingly as a place for wastes, the fishery managers must expect greater difficulties in protecting the fishes' environment.

9.6 The International Law of Fisheries

Since about 86% of the world fishery production comes from the sea, it is obviously important for the management of the fisheries to have an effective process of decision-making pertaining to the sea. Just how much of the sea fishery production comes from international waters of the high seas and how much from territorial seas is not known, but it is probable that the majority comes from the international waters.

The international fishery decisions are, of course, only a part of the problems associated with the high seas. Mineral resources, especially oil, are being exploited. Shipping plies the surface, airplanes fly above, and cables lie on the bottom. Military vehicles use the skies, surface, and subsurface areas in the interests of the security of many countries. All of these matters are involved in the balancing of the common interests in the seas, but over the recent centuries the fishery problems have generated controversy far out of proportion to their relative value.

The international law of fisheries is not a set of statutes passed by a legislative body; rather, it is a collection of principles and doctrines that has gained over the centuries a considerable degree of acceptance among nations. These principles are especially concerned with the exploitation of the resources and with their conservation. The decision makers are predominantly the officials of the several nations, occasionally judges of international courts or special tribunals. The decisions are made predominantly during the ordinary give-and-take bargaining between foreign offices and consequently are achieved to a high degree in conformity with general understanding of the doctrines.

9.6.1 SCIENTIFIC INPUTS

Prior to 1900 the fishermen from the countries of Northwest Europe had prosecuted many of the sea fisheries to the point of serious conflict over who could fish off certain coasts and to the point of serious concern about the survival of the resources. Wars had been fought over the fisheries, and numerous treaties had been developed in attempts to regulate fishing by fishermen from different countries. In addition, many countries had enacted stringent domestic laws designed to protect their fishery resources. Enough experience with the conservation laws and treaties had been obtained to generate strong feelings that the resources were not being protected and that the laws were of dubious benefit.

It was plain to many of the marine scientists of Europe that effective laws and treaties needed to be based on sound knowledge of the resources. It was plain also that the scientific problems associated with the fisheries were difficult, requiring long continued studies, and were international in scope. In 1902, the scientists organized the Conseil Permanent International pour L'Exploration de la Mer (ICES from the English title, International Council for the Exploration of the Sea). Its purpose was described in the broad statement "international investigation of the sea." Its functions were to encourage and coordinate scientific studies.

ICES has survived with great vigor to the present time while it has served to some extent as a model for other international studies. It has organized topical committees, such as Herring or Cod, and area committees, such as Baltic or Southern North Sea, according to the coherence of the scientific problems. It has published fishery statistics, hydrographic data, reports of its meetings, and a prestigious scientific journal.

ICES has played a major role in the fishery conventions concerned with the northeastern Atlantic. All of these have been negotiated to put into effect the Council's recommendations. It continues to play a key role in the coordination of scientific work being done by the several countries, fishery commissions, and international organizations.

Although ICES has had as its broad objective the investigation of the sea, it has been concerned predominantly with the northeast Atlantic. Other organizations were formed elsewhere for similar purposes, namely, the International Commission for the Scientific Exploration of the Mediterranean Sea in 1919 and the North American Council on Fishery Investigations in 1920 (discontinued in 1938).

After the formation of the Food and Agriculture Organization of the United Nations, more similar multilateral councils were organized under its sponsorship. The Indo-Pacific Fisheries Council was formed in 1948, the General Fisheries Council for the Mediterranean in 1949, the Regional Fisheries Commission for Western Africa in 1961, and the Regional Fisheries Advisory Commission for the Southwest Atlantic in 1961.

Meanwhile two bilateral conventions between Canada and the United States established the International Pacific Halibut Commission in 1924 and the International Pacific Salmon Fisheries Commission in 1937. Both Commissions were assigned the tasks of carrying out research, publishing the results and making recommendations for fishery management. Other conventions followed in many parts of the world, and they established commissions empowered, among other duties, either to carry out research or to coordinate the research among member countries. The need for international research to establish the basis for fishery management was firmly recognized.

The pertinence of research to the Law of the Sea was recognized by the United Nations in a resolution of December 14, 1954 that requested the Secretary General to convene an International Technical Conference to study the problem of international conservation of the living resources of the sea. The Conference, convened in Rome on April 18, 1955, recommended the objectives of conservation, the types of scientific information required, the types of conservation measures applicable, and reviewed the existing international conservation organizations and the major international conservation problems. Its recommendations and conclusions were reviewed and amended by a special meeting of the International Law Commission in 1956 and referred to the first major conference on the Law of the Sea.

9.6.2 CONSTITUTIVE LAW

Many of the international fishery decisions have been made on the basis of doctrines such as "freedom of the seas," and there has been little formal concensus among nations on the underlying principles. The negotiators or courts have relied frequently on "the spirit of the law" or "teachings of the most highly qualified publicists of the various nations." The closest approach to explicit principles or constitutive law was reached at the Geneva Conference in 1958.

The Geneva Conference is the colloquial name for the United Nations Conference on the Law of the Sea that convened at Geneva, Switzerland, on February 24, 1958. The conference reached agreement on four conventions that were subject to ratification by any member of the United Nations or party to the conference, and came in force after deposit of 22 ratifications. All of the conventions have come into force, but a number of important fishing nations have failed to ratify one or another of the conventions.

The conventions deal with international law of the sea in general and not merely with matters of importance to fisheries. The following are the articles considered to be of special importance to the fisheries:

Convention on the High Seas. . .

Article 1. The term "high seas" means all parts of the sea that are not included in the territorial sea or in the internal waters of a State.

Article 2. The high seas being open to all nations, no State may validly purport to subject any part of them to its sovereignty. Freedom of the high seas is exercised under the conditions laid down by these articles and by the other rules of international law. It comprises, *inter alia*, both for coastal and noncoastal States:

(1) Freedom of navigation;
(2) Freedom of fishing;
(3) Freedom to lay submarine cables and pipelines;
(4) Freedom to fly over the high seas.

These freedoms and others which are recognized by the general principles of international law, shall be exercised by all States with reasonable regard to the interests of other States in their exercise of the freedom of the high seas.

Other articles of this Convention deal especially with general commerce, safety, piracy, pollution, etc:

Convention on the Territorial Sea and the Contiguous Zone

. . . Article 1. Part 1. The sovereignty of a State extends, beyond its land territory and its internal waters, to a belt of sea adjacent to its coast, described as the territorial sea.

Article 1. Part 2. The sovereignty is exercised subject to the provisions of these articles and to other rules of international law.

Article 2. The sovereignty of a coastal State extends to the air space over the territorial sea as well as to its bed and subsoil. . . .

Article 3. Except where otherwise provided in these articles, the normal baseline for measuring the breadth of the territorial sea is the low-water line along the coast as marked on largescale charts officially recognized by the coastal State.

Article 4. Part 1. In localities where the coastline is deeply indented and cut into, or if there is a fringe of islands along the coast in its immediate vicinity, the method of straight baselines joining appropriate points may be employed in drawing the baseline from which the breadth of the territorial sea is measured.

Article 5. Part 1. Waters on the landward side of the baseline of the territorial sea form part of the internal waters of the State.

Article 24. Part 1. In a zone of the high seas contiguous to its territorial sea, the coastal State may exercise the control necessary to:

(a) Prevent infringement of its customs, fiscal, immigration or sanitary regulations within its territory or territorial sea;

(b) Punish infringement of the above regulations committed within its territorial sea.

Article 24. Part 2. The contiguous zone may not extend beyond twelve miles from the baseline from which the breadth of the territorial sea is measured.

Other articles deal especially with details of drawing baselines and with rights of innocent passage.

Convention on the Continental Shelf

Article 1. For the purpose of these articles, the term "continental shelf" is used as referring (a) to the seabed and subsoil of the submarine areas adjacent to the coast but outside the area of the territorial sea, to a depth of 200 metres or, beyond that limit, to where the depth of the superjacent waters admits of the exploitation of the natural resources of the said areas; (b) to the seabed and subsoil of similar submarine areas adjacent to the coasts of islands.

Article 2. Part 1. The coastal State exercises over the continental shelf sovereign rights for the purpose of exploring it and exploiting its natural resource.

Article 2. Part 4. The natural resources referred to in these articles consist of the mineral and other non-living resources of the seabed and subsoil together with living organisms belonging to sedentary species, that is to say, organisms which, at the harvestable stage, either are immobile on or under the seabed or are unable to move except in constant physical contact with the seabed or the subsoil.

Other articles deal especially with permanent installations and national boundary lines on the continental shelf, etc.

Convention of Fishing and Conservation of the Living Resources of the High Seas. . . .

Article 1. Part I. All States have the right for their nationals to engage in fishing on the high seas, subject (a) to their treaty obligations, (b) to the interests and rights of coastal States as provided for in this Convention and (c) to the provisions contained in the following articles concerning conservation of the living resources of the high seas.

Article 1. Part 2. All States have the duty to adopt, or to co-operate with other States in adopting, such measures for their respective nationals as may be necessary for the conservation of the living resources of the high seas.

Article 2. As employed in this Convention, the expression "conservation of the living resources of the high seas" means the aggregate of the measures rendering possible the optimum sustainable yield from those resources so as to secure a maximum supply of food and other marine products. Conservation programmes should be formulated with a view to securing in the first place a supply of food for human consumption.

Article 3. A State whose nationals are engaged in fishing any stock or stocks of fish or other living marine resources in any area of the high seas where the nationals of other States are not thus engaged shall adopt, for its own nationals, measures in that area when necessary for the purpose of the conservation of the living resources affected.

Article 4. Part 1. If the nationals of two or more States are engaged in fishing the same stock or stocks of fish or other living marine resources in any area or areas of the high seas, these States shall, at the request of any of them, enter into negotiations with a view to prescribing by agreement for their nationals the necessary measures for the conservation of the living resources affected.

Article 4. Part 2. If the States concerned do not reach agreement within twelve months, any of the parties may initiate the procedure contemplated by article 9.

Article 6. Part 1. A coastal State has a special interest in the maintenance of the productivity of the living resources in any area of the high seas adjacent to its territorial sea.

Article 6. Part 2. A coastal State is entitled to take part on an equal footing in any system of research and regulation for purposes of conservation of the living resources of the high seas in that area, even though its nationals do not carry on fishing there.

Article 7. Part 1. Having regard to the provision of paragraph 1 of article 6, any coastal State may, with a view to the maintenance of the productivity of the living resources of sea, adopt unilateral measures of conservation appropriate to any stock of fish or other marine resources in any area of the high seas adjacent to its territorial sea, provided that negotiations to that effect with the other States concerned have not led to an agreement within six months.

Article 7. Part 2. The measures which the coastal State adopts under the previous paragraph shall be valid as to other States only if the following requirements are fulfilled:

(a) That there is a need for urgent application of conservation measures in the light of the existing knowledge of the fishery.

(b) That the measures adopted are based on appropriate scientific findings;

(c) That such measures do not discriminate in form or in fact against foreign fishermen.

Other articles in this Convention deal especially with the settlement of disputes, etc.

These steps were the first small ones toward agreement on principles; many more steps must be taken. One of the most controversial matters at the Geneva Conference (as well as through the recent centuries) has been the division between the unshared authority of the coastal states and the shared authority of all states on the high seas. The region of unshared authority has been gradually widening, and now most states are willing to accept a zone up to 6 miles wide of unshared sovereignty and an additional zone out to 12 miles of unshared fisheries authority providing the coastal states recognize historic fishing rights within the 12-mile zone. The pressures to extend these zones continue, however; some countries claim 200-mile zones of unshared fisheries authority, and others are moving to draw baselines farther out or claiming lesser zones.

Significantly, no agreement was reached on the width of the territorial sea at the Geneva Conference in 1958 or at a second conference in 1960, called especially to discuss the matter.

Another especially troublesome factor affecting the fisheries is the tendency to require zones of the same width for all purposes, including customs, security, health, and fishing. Despite the absence of technical reasons for needing a uniform zone, a large number of countries and lawyers have linked a zone of fisheries authority to a zone of full sovereignty.

Furthermore, fisheries are not treated uniformly. The sedentary animal resources, such as certain mollusks and walking crabs, come under the Convention on the Continental Shelf, in which the coastal state has sovereign rights over the resources as far from shore as they are exploitable, whereas it has unshared authority over the swimming resources, fish, mammals, or some crustaceans only within a zone of a few miles wide regardless of the width of the continental shelf or the migratory habits of the species.

Such problems emphasize how little the constitutive law has progressed toward a rational management of the fisheries. There has been no recognition of technical or social goals more explicit than "conservation." There has been recognition of the need for "scientific" work but little about definitions of unit stocks or collection of the data required for their management. There has been a prevailing tendency to set boundaries of fishing zones on the basis of political factors instead of ecological ones.

9.6.3 PARTICULAR DECISIONS

Specific international controversies over fisheries (and other matters) are commonly decided by treaties or other agreements between two or more countries. Those now in force include a large number concerning the rights of fishermen, port privileges, sanitation, trade, and research, as well as the management of the fisheries. About 20 are now concerned primarily with management; to about half of these the United States is a party. Most of those concerned with management have been negotiated since 1948.

One of the early management treaties was unique in placing strong emphasis on research and in granting regulatory powers to the Commission that it created. This treaty, between Canada and the United States, established the International Fisheries Commission (now called the International Pacific Halibut Commission), which has been strikingly successful in restoring the abundance of halibut and increasing the annual catch by more than 50%. This success and the means by which it was achieved have served as an example to other people confronted with declining fishery resources.

The halibut fishery off the west coast of North America began in 1888 and found a large market in the Middle West soon after the completion of the transcontinental railroads. The halibut were caught by line gear set on the bottom along the coast up to about 100 miles from shore from off the northern coast of California north to the Bering Sea.

The fishery went through the now familiar cycle of early profits and then depression. The halibut on the first grounds fished, off Vancouver Island and Washington, soon declined in abundance; and the fishermen moved north to off Southeastern Alaska and then west to off the Alaska Peninsula. The total catch increased to 31,000 tons in 1915 and then declined despite heavier fishing.

Some concern had been felt long before the peak, and the two countries included discussions of conservation matters along with their negotiation of tariffs and port privileges, but they reached no agreement on conservation until they ratified the Halibut Convention of 1923 in 1924. This Convention provided for a winter closed season and the establishment of a Commission that was to make a thorough investigation of the life of the halibut and was to recommend regulations. Regulatory authority was conferred upon the Commission in 1930.

The facts necessary for appropriate regulation as well as acceptance of regulation by the fishermen were established by the scientific investigators. First came evidence that the decline was not just a natural fluctuation but was a result of fishing. This was provided (1) by studies of the life expectancy, which showed the halibut to be a very slowly growing fish; (2) by close correlation of declines in local populations with the amount of fishing; and (3) by evidence in the age composition of reasonably consistent recruitment of young. Next was definition of the stocks. They were studied by tagging, by estimation of the drift of eggs and larvae, by comparison of morphometry, and by comparison of rate of growth. It was found that intermingling was complex since eggs and larvae drift for long distances, immature halibut migrate little, and mature halibut migrate extensively. A point of substantial separation was found at Cape Spencer, Alaska, but evidence of less-separated units was found elsewhere. A decision was made to regulate on the basis of stock in two principal areas; Area 2, from Willapa Bay, Washington, to Cape Spencer; and Area 3, from Cape Spencer to the Aleutian Islands.

Coincident with these studies were a full collection of historical data on the catch and effort of the fishery and a study of the effect of fishing on catch. It was found that there had been a 65–80% decline in the catch per unit of effort (the index of the size of the stock) associated with the increase in fishing. The rates of fishing together with natural mortality were estimated to be higher than the rates of recruitment and growth. There was good evidence that both the size of the stock and the yield could be increased by a reduction in the amount of fishing.

The Commission reported its findings in full to the governments of Canada and the United States and recommended annual catch quotas on each of the two defined stocks. They also recommended continuation of the winter closed season for prevention of fishing on the schooled spawners, protection of certain nursery areas, and prohibition of catching the young halibut while they were rapidly growing.

The quotas were set for the first time in 1932 and each year thereafter at levels a little below the estimated replacement by reproduction and growth. This practice allowed the fishery to continue while the stock size and yield were gradually increased. It worked well. The total annual yield, which had declined to about 20,000 tons in 1930 after the peak of 31,000 tons in 1915 rose to a sustained level of near 30,000 tons in 1960. The size of the stocks in 1960 was more than twice as large as in 1930, the total amount of fishing only half as much. Good scientific evidence indicated that the yield had risen to near the maximum sustainable level.

The relative economic success of the average fisherman and investor did not follow, however, the success in restoration of the resource. The fishermen had overinvested in boats during the expansion of the fishery and were still operating about 400 when quota regulation began in 1932. Despite the quotas and a fleet already too large, the number of vessels increased to more than 500 in 1937 and more than 800 in 1950. Since the total amount of fishing needed to catch a quota was being reduced, fishing seasons became much shorter; in some years less than 1 month in one area and less than 2 months in the other. A full explanation of the economic reasons for this shift of vessels into the halibut fishery has not been made, but it was probably due to the relatively low earnings and income on investments in other kinds of fishing. The consequences were, however, an antiquated fleet and low levels of earnings for halibut fishermen or investors at the time the halibut stocks were fully restored.

Another factor related to the regulatory practices of the Commission was a decline in prices to the fishermen as a result of increased cost of marketing. Prior to regulation, the landings were spread over much of the year and mostly marketed fresh, but after regulation, the landings were concentrated during the shortened seasons, when they occasionally glutted the market and required much larger freezing and storage facilities for proper marketing.

Then, during the 1960's the halibut stocks that had been so painstakingly restored over 30 years became endangered again. The halibut are, of course, only a small fraction of the biomass of fish on the continental shelf of the Gulf of Alaska. Other species, especially rockfish, flounders, walleye pollock, and hake were sought by Japanese and Russian trawl fleets. Inevitably, these trawlers included a small fraction of halibut in their catches despite the fact that they did not seek halibut especially. The consequence was considerable catches of halibut not under control of the Halibut Commission and possibly large enough to endanger the resource again. The problem had shifted to one of the best management strategy for the combined fisheries.

9.7 Special Problems of Fishery Development

People are not always satisfied with a passive management of the fishery resources and want to accelerate their development. Such a development is part of a broader economic development, and in most of the less-advanced countries, the development of food resources must precede the development of diversified industry. The fishery resources may be relatively much less important in the more-advanced countries, but even so, they can be considered to be in a process of continuous development. In this section major problems associated with various kinds of fishery development will be identified.

The problems of fishery development span a broad range of technical, economic, and political questions, but they derive from our basic desire for economic growth, modified by a concern for living things and the environment. The major objectives are to supply food or outdoor recreation at minimal cost, to increasing numbers of people on a sustainable basis, and at the same time to integrate the needs of fisheries with the needs of other users of water. The attainment of these objectives requires the formulation of policies in which objectives and general principles are identified in the preparation of detailed plans.

The policy-making and planning process involves two major steps. The first step is the collection of basic data on resources, facilities, skills, markets, and institutions and identification of possible routes of development. The next is the identification of the pertinent decisions, decision makers, and methods of improving the decisions for accomplishment of the objectives.

The resources, institutions, technical alternatives, and stage of development vary greatly among the different types of fisheries, and hence it is desirable to discuss them separately.

9.7.1 INTERNATIONAL FOOD FISHERIES

The fisheries outside the territorial seas and fishing zones of coastal nations are conducted predominantly by expensive, technically sophisticated vessels that can operate a long distance from port. The vessels are owned by private, profit-oriented operators, who are frequently subsidized by governments or by agencies of centrally planned governments. There is no aquaculture in this area now nor is there likely to be any in the foreseeable future.

Few fishery resources in this area are subject to any institutional control. Almost all are available without restriction to fishermen who are able to find them and catch them. Such resources have been especially attractive to large vessel operators who have been able to take quick profits from one stock and move on to another and then to another as the abundance has declined in each. The operators and their governments adhere strongly to the principle of the freedom of the seas, which allows them to fish anywhere in international waters and commonly reject responsibility for maintenance of the resources.

The most urgent problem in the development of these fisheries is likely to be the formation of effective controls that will restrain overexploitation of the stocks without inhibition of the discovery of new stocks or the innovative creation of a new fishing methods. This problem can be approached through a central international authority, regional bilateral or multilateral agreements, or extension of the authority of coastal states. The latter two seem to be the most probable routes at this time although no extension of coastal authority now contemplated would include all ocean fisheries.

Regardless of the approach, there is a major need for much better information on the resources, the levels of sustainable yields, and the consequences of alternate strategies of management.

9.7.2 NATIONAL FOOD FISHERIES

The fisheries in inland waters, territorial seas, and contiguous fishing zones are pursued by a mixture of large, sophisticated and relatively small, primitive equipment that is all profit-oriented or operated by agencies of centrally controlled governments. The large equipment may be also used to fish in the international areas. Aquaculture is a substantial and rapidly growing part of these fisheries.

The mixture of large, sophisticated and small, simple equipment in the national fisheries poses a different set of political problems than those of the more sophisticated, longer-distance operations of the international fisheries. The small, simple equipment is usually operated by large numbers of traditional fishermen who cannot easily change their practices because of lack of skills, or lack of money, or distance from markets. The encouragement of efficient lower-cost fishing by fewer fishermen may produce secondary problems of unemployment and the need to retrain or relocate the displaced fishermen.

Another contrast with the international fisheries lies in the national concern for the welfare of the national resources. These are plainly the responsibility of the government, and their future yields depend on the present course of management.

The concern for both fishermen and the resource creates a strong political pressure toward inefficiency that opposes the economic pressures developed by competition with other foods and the rapidly increasing international trade in fishery products. The fishermen find increasing difficulty in keeping costs down, especially if they depend on stocks whose yields have reached or declined below maximum sustainable levels because of heavy fishing. In such fisheries, the levels of earnings and return on capital are both too low. Development of such inefficient fisheries depends primarily on raising the efficiency through controlled reduction of the number of fishermen and amount of equipment if private investment is to have a part.

Other natural stocks, not as heavily exploited, may offer opportunities for development if yield, processing facilities, and markets can be expanded. The government can do much to help the private sector by surveying the resources and estimating the levels of potential sustainable yields. It may also help through planning, construction, and operation of demonstration vessels equipped with special gear, or the construction of demonstration processing plants. It may help to train fishermen, fishery extension agents, scientists or processing plant supervisors. It may set and enforce sanitary standards for the operations and quality standards for the products. The functions and the priority given to the alternatives must be decided on the basis of intimate knowledge of the local operations and their weak points, but above all, it must solve the basic conflict between protecting the resources and allowing an attractive return to labor and capital.

9.7.3   RECREATIONAL DEVELOPMENT

The recreational activities associated with fisheries include recreational or sport fishing, which is provided predominantly by governments; fish watching at aquaria, fish hatcheries, or fish activity areas, which is provided about equally by government and private means; and keeping fish as pets, which is provided almost wholly by private means. The development of recreational fishing is of primary concern.

Recreational fishing as well as public activities suitable for fish watching are parts of a broader outdoor recreational experience involving travel, living outdoors, hiking, and boating. Fishing is an essential but perhaps only a minor part of an activity that may be highly varied. Indeed, the variety of experiences is one of the attractions for many people.

The development of varied recreational fishing opportunities for more people is intimately associated with the maintenance and improvement of the outdoor environment. It can be expected to involve control of pollution, development of varied transportation systems from highways to hiking trails, provision of access to fishing areas, management of new waters, as well as better management of existing waters.

Governments must play a major role in the planning and execution of all of these factors. Their responsibility starts with the fishery resources that are almost entirely public. These must be preserved and their use divided between commercial and recreational fishermen as well as among the recreational fishermen. Next is the control of the environmental quality, which must be accomplished largely through government control of private activities. Next is the management of new waters, usually reservoirs, for fishing in combination with other uses, provision of more fish or more species for stocking, and provision of highways and access opportunities. Inherent in many of these responsibilities is the need to provide for the fisheries in the face of demands for other uses of the water. Inherent also is the need to plan adequately for the protection of resources before damage occurs and for the provision of opportunities at the time they are needed.

The private, profit-motivated activities related to recreational fisheries are mostly of a supporting type: provision of equipment, lodging, and transportation. The expenditures for these purposes are a large fraction of the amounts spent by sport fishermen and provide attractive opportunities for private investment. Other opportunities are appearing in the management of waters for angling or the sale of fish from private farms to public agencies.

The development of recreational fishing is a response to an emotional demand by a large and increasing proportion of the population. An adequate response will require extensive strengthening of the government institutions responsible for the management of both the waters and the fisheries.

REFERENCES

No clear separation is possible between the references on fishery management and those on exploited populations listed at the end of Chapter 6. The latter should be consulted in addition to those listed below.

Bennett, G. W. (1962). "Management of Artificial Lakes and Ponds." Van Nostrand-Reinhold, New York.

Benson, N. G., ed. (1970). "A Century of fisheries in North America," Spec. Publ. No. 7, pp. 1–330. Amer. Fish. Soc., Washington, D. C.

Christy, F. T., and Scott, A. (1965). "The Common Wealth in Ocean Fisheries." Johns Hopkins Press, Baltimore, Maryland.

Crutchfield, J. A. (1962). Valuation of fishery resources. *Land Econ.* **38**, 145–154.

Gilbert, DeW., ed. (1968). The future of the fishing industry of the United States. *Univ. Wash. Publ. Fish.* [N. S.] **4**, 1–346.

Gulland, J. A., and Carroz, J. E. (1968). Management of fishery resources. *Advan. Mar. Biol.* **6**, 1–71.

Hamlisch, R. (1962). Economic effects of fishing regulation. *FAO Fish. Rep.* **5**, 1–599.

Johnston, D. M. (1965). "The International Law of Fisheries." Yale Univ. Press, New Haven, Connecticut.

McKee, J. E., and Wolf, H. eds. (1963). "Water Quality Criteria," 2nd ed., Publ. 3–A. Resources Agency of California, Sacramento, California.

Polikarpov, G. G. 1966. "Radioecology of Aquatic Organisms." North-Holland Publ., Amsterdam.

United Nations, (1955). "Report of the International Technical Conference on the Conservation of the Living Resources of the Sea," United Nations A/Conf, 10/5 Rev. 2. United Nations.

New York. "United Nations. (1956). "Papers Presented at the International Technical Conference on the Conservation of the Living Resources of the Sea," United Nations A/Conf. 10/7. United Nations, New York.

Vibert, R., and Lagler, K. F. (1961). "Pêches continentales, biologie et amenagement." Dunod, Paris.

Subject Index